未来への共創
Co-innovating tomorrow

横河電機が挑んだ
リブランディングの軌跡

Co-creation for the Future.
The Challenge of Rebranding Yokogawa

編著：横河電機ブランドブック制作委員会
発行：日経BP
発売：日経BPマーケティング
Written and edited by
Yokogawa Electric Corporation Brand Book Production Committee
Published by Nikkei Business Publications, Inc.
Released by Nikkei BP Marketing, Inc.

未来をともに……

What's Next?

CONTENTS

ブランディングの背景
Background of Brand Rebuilding

横河電機は、その前身である電気計器研究所が1915年に創設されて以来、計測・制御事業全般にビジネスを拡大し、日本のモノづくりを代表するブランドとして産業界を支えてきた。

しかし、産業界を取り巻く変化のスピードの加速など様々な環境の変化の中、従来のビジネスモデルにとどまっていては将来の継続的な成長を見込むことは難しい。そこで、横河電機が2000年代に入ってから取り組んだのが、企業としてのアイデンティティの整理とそこに立脚したブランドの再構築である。

中核事業である制御事業における提供価値を表現したVigilance、製品やサービスと結びついたお客様の使用価値を表現したVigilantPlantというビジネスコンセプトを立ち上げることで、海外を含めた事業は順調に拡大していった。

2010年代に入り、様々な製品やサービスがネットワークでつながる時代となると、横河電機は新しい時代に対応すべく、顧客との共創をベースにしたソリューション提供企業を目指す方針を打ち出した。

そこで必要とされたのが、製品やサービスなどの事業レベルを超えたブランド再構築を含む、コーポレート全体のブランディングだ。

折しも、2015年に創立100周年を控えており、そのタイミングで新しい長期経営構想、中期経営計画が練られていた。まさに、メーカーとしての横河電機から、ソリューション提供企業としてのYOKOGAWAグループへの進化を加速する動きとリンクして、ブランディング推進チームによる活動が本格化したのである。

Following its founding in 1915 as an electric meter research institute, Yokogawa Electric Corporation expanded its business to cover all areas in the measurement and control fields, and supported industry as a leading brand in Japan's manufacturing sector.

Nevertheless, in a rapidly changing business environment, simply staying with conventional business models offered uncertain prospects for sustained growth. Therefore, in the 2000s, Yokogawa began to take a fresh look at its corporate identity and retool its brand based on changes made in this identity.

It steadily expanded its business around the world by introducing the Vigilance concept, which articulated the value proposition of Yokogawa's core industrial automation and control business, and the VigilantPlant business concept, which made clear the value that customers could derive from using Yokogawa's products and services.

As the connection of various products and services over networks gathered pace in the 2010s, Yokogawa made it a policy to engage with its customers in the co-creation of solutions that met the needs of a new era.

This required a comprehensive approach to corporate branding and a restructuring of the brand that went beyond the product and service business level.

Coinciding with preparations to mark Yokogawa's 100th anniversary in 2015, the opportunity was taken to create a new long-term business framework and mid-term business plan. The Branding Promotion Team linked its activities with an initiative to accelerate the company's evolution from manufacturer to a global group of companies engaged in the provision of solutions, and shifted these activities into high gear.

新たな事業展開を実現するためのブランディング

Branding for the Development of New Businesses

ブランドとは、顧客や社会の間に築かれる評判のことである。社外への宣伝・広報活動だけでなく、提供する製品やサービス、個々の社員の言動などもまた企業の評判に関わる要素であり、それらすべてがブランドを構成する要素だと言える。

特に、横河電機が取り組んできたBtoBビジネスのブランディングでは、顧客企業の経営者や担当者を含めた複数の意思決定者、さらには市場全体を見渡して、目標とする事業展開に沿ったブランドを確立することが、ビジネスを展開していく上で極めて重要なポイントとなる。

経営陣が方針を決定したからといって、真のソリューション提供企業に変化することは容易ではない。顧客や市場に新しい姿を提示して、企業に対する認識を新たにしてもらう必要がある。それ以上に、企業で働く社員が方針を受け入れると同時に、自らが企業文化を変革し、進んで外部へ発信することが欠かせない。ブランディング推進チームは、まずそのよりどころとなる指標や道筋を示すことに取り組んだ。

Brands are built on a reputation that is established with customers and the general public. Key considerations include not only external advertising and publicity activities but also the actual products and services as well as the words and actions of individual employees. These are the elements on which every brand is based.

Particularly with the BtoB branding that Yokogawa is engaged in, the establishment of a brand for the development of a target business must take into consideration multiple decision-makers at all levels of customer organizations as well as the broader market. This is an incredibly important aspect of the development of such businesses.

But just because a management team has decided on a policy of becoming a solutions provider does not mean that this transformation is easy to achieve. A new business model must be presented to customers and the markets to reestablish recognition of the brand. Even more essential is gaining the acceptance of employees, while at the same time transforming the company's culture from within. The Branding Promotion Team's first step in this process was to establish KPIs and a roadmap to serve as a foundation for these activities.

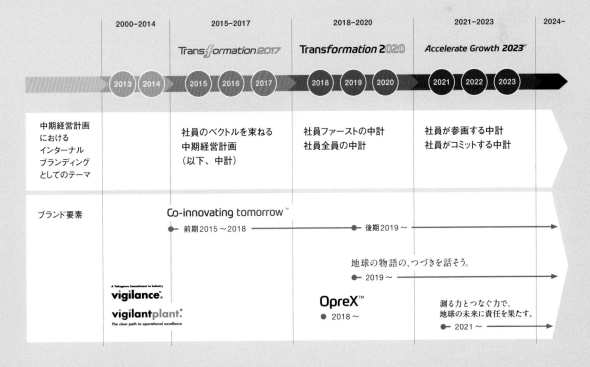

横河電機によるブランディングの取り組みとその変遷

Yokogawa's branding efforts and developments

共創をキーワードに活動を展開
Activities Based on the Keyword "Co-creation"

活動の最初の大きな成果が、創立100周年の記念日に合わせて発表されたコーポレート・ブランド・スローガン「Co-innovating tomorrow」である。企業としてのあるべき姿を顧客や社員に示すために端的な言葉で表現したメッセージで、明日に向かって、様々な立場の人たちと対話しながら、新しい価値を「共」に「創」り上げ、イノベーションを共に巻き起こしていくという強い意志は、YOKOGAWAグループのブランディングのキーワードでもある。

　次いで、このスローガンに込めた想いをビジュアルで訴求するキー・デザイン・エレメントとして、顧客から評価されている安定性や技術の正確性、サステナビリティ、顧客との共創を象徴した「Brilliant Grid」(ブリリアントグリッド)を作成した。以後、Co-innovating tomorrowとBrilliant Gridを核とした広告やグラフィックが展開されていく。

　コーポレートブランドが確立したところで、事業レベルでのブランディングとして、製品ブランドの体系整理も進められた。それまでYOKOGAWAグループ全体で1000以上も存在した制御事業に関連する製品ブランドは新たに立ち上げた「OpreX」(オプレックス)に統合され、すべての製品とサービスおよびソリューションなどは、OpreXの名の下に分類されることになった。

The first major accomplishment of these activities was the announcement of the "Co-innovating tomorrow" corporate brand slogan on the occasion of Yokogawa's 100th anniversary. This succinct message was intended to convey the company's aspirations to customers and employees, and is a key phrase in the Yokogawa Group's branding that communicates a commitment to creating new value and innovating while engaging with stakeholders.

Next, the Brilliant Grid, a key design element that symbolized the stability, technical precision, sustainability, and co-creation with customers that Yokogawa is known for, was created as a visual expression of the meaning behind the corporate brand slogan. Both "Co-innovating tomorrow" and the Brilliant Grid became core elements in Yokogawa's advertising and graphics.

Once the corporate brand was established, streamlining of the product brand ecosystem was undertaken as a branding activity at the business level. Product brands for more than 1,000 industrial automation and control businesses across the entire Yokogawa Group were unified under the newly launched OpreX brand, and all products, services, and solutions were classified into OpreX categories.

Co-innovating tomorrowとBrilliant Gridを使ったクリエイティブ

Creative using "Co-innovating tomorrow" and the Brilliant Grid

横河電機における制御事業を包括するブランドとして立ち上げた「OpreX」

OpreX, the comprehensive brand launched for Yokogawa's industrial automation and control business

社員全員が参画したパーパスの策定

Creation of the Purpose Statement with the Full Participation of Employees

YOKOGAWAグループのブランディングが成功した要因の一つとして、経営陣や社員にブランドに対する考え方が浸透し、社員がブランドアンバサダーとしての自覚をもって関与したことが挙げられる。イントラネットでの情報共有は、そのための重要な手段であった。各種ガイドラインやクリエイティブ向けのテンプレートを集約するだけでなく、コンテンツを継続的に追加して情報を発信し続けることで、ブランディングが一時的なキャンペーンではなく、継続的に推進すべき取り組みであるとの意識を浸透させることができた。

ブランディングの新しい方向性が見えてきたのは、2019年の「地球の物語の、つづきを話そう。」をキャンペーンタグラインとした広告出稿と、同名のブランド・キャンペーン・サイトの公開だろう。地球環境やサステナビリティを意識したコンテンツを発信していくことで、YOKOGAWAグループの新しい姿を、顧客や社員のみならず広く一般に示すことができた。特に、社員の熱い反応にブランディング推進チームのメンバーは大きな手応えを感じ、それが「社員が参画しコミットするパーパス」につながっていく。

One of the factors contributing to the success of the Yokogawa Group's branding initiative was the broad appreciation of the brand concepts by the management team and employees, and the employees' dedicated involvement in activities undertaken with the mindset of a brand ambassador. Sharing information over the company intranet was a key means for achieving this. Yokogawa used the intranet not only as a repository for guidelines and creative templates, but also to foster awareness that branding was more than a short-lived campaign and was something that should be promoted on an ongoing basis. It did this by continually adding new content and communicating information about its brand.

Yokogawa's branding took off in a new direction with the placement of ads in 2019 that used the campaign tagline "What's next for our planet? Let's make it smarter." and the launch of a brand campaign website with the same tagline. The release of content focusing on the environment and sustainability widened awareness of the Yokogawa Group's new model among customers and employees as well as the general public. The Branding Promotion Team knew they had hit upon something when they saw how this message resonated with Yokogawa's employees, and contributed to their participation in the creation of the Purpose Statement.

「地球の物語の、つづきを話そう。」広告クリエイティブ

"What's next for our planet? Let's make it smarter." ad creative

2020年代に入り、世界情勢の流動化やコロナ禍も相まって、企業を取り巻く社会環境は激変していった。そうした困難を乗り越えるには、企業としての社会的な使命と存在意義を端的に表すパーパスが必要であると考えた経営層からの要請を受け、ブランディング推進チームは社員全員が参画するパーパス策定を実現した。

こうして策定されたパーパスが、「測る力とつなぐ力で、地球の未来に責任を果たす。」である。社員の熱意とコミットメントがこもったこのパーパスは、YOKOGAWAグループ全社員の精神的な支柱として、今後長く使われていく。

次章から5つのパートに分けて、ブランディング推進チームが中心となって取り組んできた横河電機のブランド再構築プロジェクトの軌跡をたどる。グローバルにBtoBビジネスを展開する日本企業が、どのような意志や目的を持ってブランディングに取り組み、どのような成果を得てきたのか。さらに、活動を続けていく「明日」の先に何を見いだそうとしているのかを探っていく。

Then at the start of the 2020s, Yokogawa found itself in a drastically changed situation due to a dramatic shift in world affairs and the onset of the COVID-19 pandemic. To overcome these adverse conditions, the management team thought that needed a purpose statement that succinctly expressed its social mission and reason for existence. Acting on this request, the Branding Promotion Team created a purpose statement with the full participation of their fellow employees.

The Purpose Statement that the team came up with reads, "Utilizing our ability to measure and connect, we fulfill our responsibilities for the future of our planet." Embodying employees' passion and commitment, this statement will be used long into the future and it will be something that all Yokogawa Group employees can turn to for guidance.

The next five chapters trace events and developments in the Yokogawa brand reestablishment project led by the Branding Promotion Team. What were the aspirations and objectives behind the branding efforts of a Japanese company operating a global BtoB business, and what were the results? The following chapters explore what comes next for this company as it continues its activities.

測る力とつなぐ力で、地球の未来に責任を果たす。

Utilizing our ability to measure and connect, we fulfill our responsibilities for the future of our planet.

2021年5月に中期経営計画の発表に合わせて発表されたYokogawa's Purpose

Yokogawa's Purpose unveiled together with the release of the mid-term business plan in May 2021

Chapter 1

契機
A Great Chance

創立100周年を前にして
動き出したブランドの再構築

Reestablishment of the Brand in Preparation for the 100th
Anniversary of Yokogawa Electric Corporation

ブランド再構築への動き
Background of the Corporate Brand Reestablishment Project

　横河電機のブランド再構築プロジェクトに新しい動きが本格的に始まったのは、創立100周年を2年後に控えた2013年だった。

　2003年から取り組み始めた活動で、VigilanceやVigilantPlantという制御事業のビジネスコンセプトは既存顧客の中で定着し、YOKOGAWAの認知度も海外で高まるなど、当時のブランド活動の成果は一定の評価を得ていた。

　しかし、これらの活動からすでに10年の歳月が経過しており、横河電機を取り巻く環境は大きく変化した。IT化が急速に進み、様々な製品やシステムがネットワークでつながったことで、顧客は企業経営全体を見渡したコンサルティングやソリューションを求めるようになり、計測機器や制御システムを提供するだけでは、付加価値の高いサービスを提供できなくなりつつあったのだ。

　VigilanceやVigilantPlantは、あくまで制御事業のビジネスコンセプトに過ぎない。変化の激しい時代において競合するIT企業を含む世界の巨大企業と戦うには、従来のブランド戦略のままでは取り残されてしまう。2013年当時の横河電機にとって、10年前に構築したブランドの再構築は待ったなしの局面を迎えていた。こうした状況を背景にして、横河電機のブランディング推進チームは活動を開始したのである。

It was in 2013, two years before the 100th anniversary, when Yokogawa Electric Corporation began to move ahead with its brand reestablishment project.

Yokogawa's branding activities started in 2003 and achieved positive results based on industrial automation and control business concepts such as Vigilance and VigilantPlant that targeted existing customers and increased recognition of the company overseas.

Ten years following the initiation of these activities, the environment surrounding the company had changed significantly. IT was being rapidly implemented, and a wide range of products and systems were being connected via networks. This increased customer demand for consultation and solutions related to overall corporate management, and it became nearly impossible for the company to provide high-value-added services solely through the provision of existing measurement devices and control systems.

Vigilance and VigilantPlant were only industrial automation and control business concepts, and the company's existing branding strategies were insufficient to compete with large global corporations, including IT companies, in a dynamically changing business environment. In 2013, it became urgently clear that Yokogawa needed to reestablish its brand. With this purpose in mind, the Yokogawa Branding Promotion Team launched its activities.

企業文化を変革する装置としてのブランディング
Branding as an Apparatus for Transforming Corporate Culture

2022年度のグループ全体の海外売上高比率が70%を超える横河電機は、グローバル企業と呼ぶにふさわしい存在だと言える。海外売上高比率は2013年当時もすでに65%を超え、企業文化や体質はグローバル化の方向に進みつつあったものの、まだまだ多くの課題が存在していた。

例えば、事業部間の横のつながりが薄く、情報が十分に共有できていなかったため、異なる事業部から類似した製品が提供されることがあった。また、日本と海外拠点とのコミュニケーションも十分とは言えなかった。この状態が続けば、市場や顧客の混乱を招きかねず、ひいては長年培ってきたお客様との信頼関係にも影響を与える恐れもあった。

かつては事業部ごとに製品をつくり、互いに切磋琢磨し合える時代もあったが、複数の製品やサービスがネットワークでつながる時代では、事業部間のコミュニケーションが取れていなければ、顧客に信頼されるソリューションを提供できない。当時の横河電機には、事業横断的な戦略が求められていたのだ。

ではどうすればよいのか。ブランディング推進チームが検討したのが、"ブランドを活用した企業文化の変革"であった。
「新しい時代に適応できるように企業文化や体質を変えることが、ブランディングという手段を利用することでできるのではないかと考えました。いわば、企業文化を変革する装置として活用するわけです」とメンバーの一人は回想する。

横河電機におけるブランド再構築の目的は、単に顧客に対する認知度向上によって競合する巨大グローバル企業と戦うためだけでなく、企業文化自体を変革して横河電機を真のグローバル企業に成長させることへと変化していったのである。

With an overseas sales ratio that exceeded 70% in fiscal year 2022, Yokogawa can be said to be a global company. Although the overseas sales ratio in 2013 had already exceeded 65%, there were still many aspects of the company's corporate culture and character that had not yet been globalized.

For example, a lack of communication among its business units led to the insufficient sharing of information, with the result that different business units ended up providing similar products. Communication between Japan and Group companies in other countries was also inadequate. Continuing to do things the same way would have led to confusion in the market and with customers, which would have had an unfavorable impact on the long-lasting trust-based relationships that we had developed with customers.

In the past, individual business units manufactured their own products and improved through friendly rivalry with one another. But now that multiple products and services were being connected via networks, it was proving impossible to provide solutions that customers could trust without there being good communication between business units. Yokogawa recognized the need for the establishment of cross-sectoral strategies.

The Branding Promotion Team discussed the issue of pursuing a brand-driven transformation of Yokogawa's corporate culture.

"With branding, we were confident that we could adjust to a new era and change the company's corporate culture and characteristics. In other words, we thought it would be a good idea to use our brand as an apparatus for transforming our corporate culture," said one of the members of the team.

The purpose of Yokogawa's brand reestablishment shifted from gaining customer recognition to compete with large global companies to growing into a real global company through the transformation of its corporate culture.

グローバル・ブランド・マネジメント体制の確立

Establishment of a Global Brand Management Structure

ブランド再構築プロジェクトの中心になったのは、マーケティング本部である。マーケティングコミュニケーション統括部門の下、ブランディング推進チームを中心メンバーに、社内コミュニケーション、PR、デジタルマーケティングの各機能が一体となって、新たなブランディングを行っていく体制が整えられた。

さらに、プロフェッショナルなブランディングを実現するため、戦略、クリエイティブ、テクノロジーを組み合わせてブランドとビジネス双方の成長を促進する支援を行うブランディング専門会社とも協業していくこととなった。

ポイントとなったのは、世界各拠点でコミュニケーション業務を担うGMT（Global Marketing communication Team）を巻き込んだ活動だ。その理由についてブランディング推進チームの一人は、「本社機能が日本にある企業はどうしても日本中心に物事を考えがちです。横河電機のようにグローバル展開している企業にとって、海外拠点のチームとのコミュニケーションがブランディング成功の鍵となるのです」と語る。

ブランディング推進チームは、海外メンバーを含むGMTとの意見交換を通して、信頼関係を築きながら活動を進めていった。各期に1度、GMTメンバー全員が集まるオンラインミーティングを実施し、年に1度は、全員が、三鷹の本社に集まる対面ミーティングも行った。当初、GMTメンバーは本社のガイドをフォローする存在であったが、回を重ねるごとに徐々に積極的に参画する姿勢を見せるようになり、互いの活動を共有し、横展開を図るようになっていった。こうした取り組みにより、グローバル・ブランド・マネジメント体制を確立していった。

The Marketing Headquarters played a central role in the brand reestablishment project. Centering on the Branding Promotion Team under the Marketing Communications Department, the in-house communication, PR, and digital marketing sections worked together to create a new branding style.

Furthermore, to realize professional branding, Yokogawa also decided to cooperate with a branding company that supported the promotion of both brand and business growth through a combination of strategies, creativity, and technologies.

One of the activities that had a significant impact on the project involved the Global Marketing Communication Team (GMT), which is responsible for communication activities at Group companies around the world. A member of the Branding Promotion Team said, "Companies with their headquarters functions in Japan tend to have a Japan-centric mindset. For companies like Yokogawa that are actively expanding their business around the world, communication with their overseas teams is key to success in branding."

Through the exchange of opinions and other means, the Branding Promotion Team sought to build good relationships with GMT members around the globe. On a quarterly basis, all GMT members met online, and everyone came once a year to the headquarters in Mitaka City, Tokyo, for face-to-face meetings. Initially, the GMT members merely followed the guidance provided by the headquarters; however, they gradually began to actively participate in branding activities by sharing individual approaches and working together. Through these approaches, they established a global brand management structure.

グローバル・ブランド・マネジメント体制（本社組織）

Global brand management structure (headquarters organization)

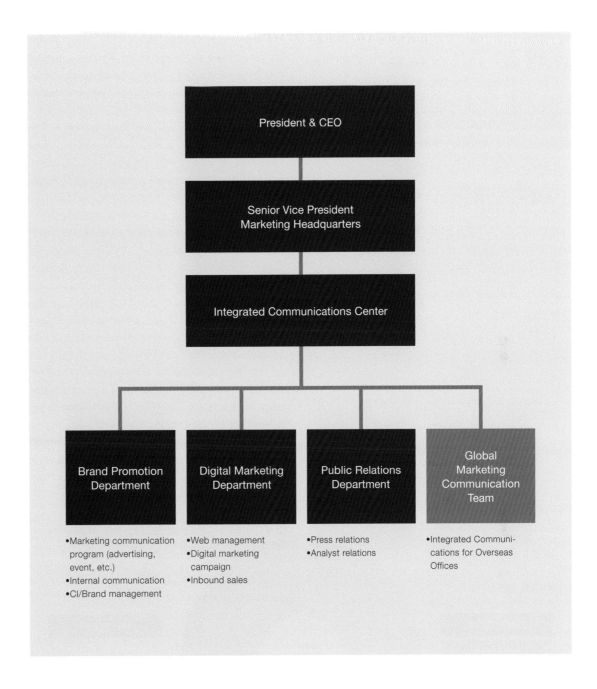

President & CEO

Senior Vice President
Marketing Headquarters

Integrated Communications Center

Brand Promotion Department	Digital Marketing Department	Public Relations Department	Global Marketing Communication Team
•Marketing communication program (advertising, event, etc.) •Internal communication •CI/Brand management	•Web management •Digital marketing campaign •Inbound sales	•Press relations •Analyst relations	•Integrated Communications for Overseas Offices

現状整理のための3つのアプローチ
Three Approaches to Analyzing the Current Situation

ブランディング推進チームがまず取り組んだのは、現状の整理であった。現状を正確に把握して初めてブランディングの方向性を見いだせると考えたからだ。

具体的には、「①市場からどう見られているのか」「②企業としての明確な意思は何か」を明らかにし、その結果をもとにして「③差別化のための磨くべきコアコンピタンスは何か」を見つけるという手順をとった。

①市場からどう見られているのか

2013年10〜11月、既存の顧客を対象にYOKOGAWAのブランドイメージ調査を実施した。市場におけるイメージ、顧客の評価、競合他社との違いについて、日本および海外9拠点で約1300件のデータを収集し、分析した。

その結果、ブランドの特長として評価されている点が

The Branding Promotion Team first set about analyzing Yokogawa's current situation. This was because they believed that they would only be able to determine a direction for the brand after they had gained an accurate understanding of this.

What they did specifically was 1) identify how the markets view Yokogawa, 2) clarify what its intentions were as a company, and based on those findings, 3) identify what core competencies it should enhance to differentiate itself.

1) How Do the Markets View Yokogawa?

In October and November of 2013, the team conducted a brand image survey of existing customers. They collected and analyzed approximately 1,300 pieces of data from Japan and nine overseas locations on Yokogawa's image in the markets, its reputation among customers, and how it differed from its competitors.

The results revealed brand strengths in responsiveness and quality; but when the team created a brand positioning map that included other companies, it was found that VigilantPlant was not as strongly associated with responsiveness and quality as the Yokogawa brand. In short, impressions of VigilantPlant and the Yokogawa brand diverged sig-

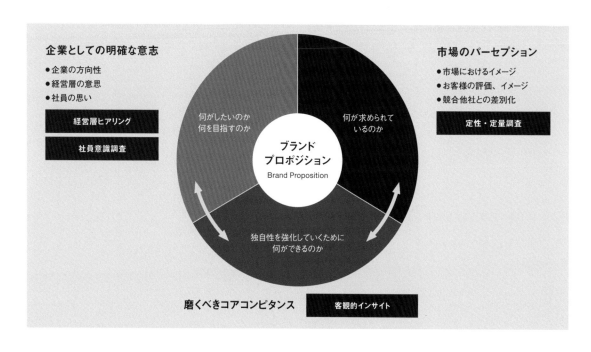

現状整理のためのアプローチ

Approaches to analyzing the current state

「対応力」や「クオリティ」であることが判明したが、他社も含めたイメージマップを作成してみると、VigilantPlantはYOKOGAWAブランドほど、「対応力」や「クオリティ」をイメージさせるものではないという結果が出た。つまり、2つのブランドイメージには、大きな乖離があったのだ。ブランド再構築の必要性が改めて明らかになった。

②企業としての明確な意思は何か

次に、経営層へのヒアリングと社員の意識調査を通じて、企業の方向性、経営層の意思、社員の思いを確認した。ここで重要なのは、経営層だけでなく社員が、「どうあるべきか、どうありたいか」という存在意義を意識しているか、そしてそれが行動に落とし込まれているかの確認である。特に海外の社員からは、「もっと新鮮で革新的なイメージが欲しい」「新しいブランドビジョンが必要」といった声が寄せられた。

こうした結果から、製品メーカーにとどまらず、顧客のビジネスパートナーとしてソリューションを提案する企業への変革が必要であると分かった。

③差別化のための磨くべきコアコンピタンスは何か

①②の結果をもとに、YOKOGAWAグループが独自性を強化していくためには何をすべきかを、ベンチマークとした競合とも比較しながら検討した。そこでの結論は、これまで培ってきた「高品質」や「遂行力」といった顧客が持つイメージを維持しながら、ソリューションやビジネスパートナーとしての力を磨き、「革新性」や「リーダーシップ」といったイメージを形成することが、長期的な競争優位に結びつくというものだった。

さらに、創立100周年事業との統合、後述する長期経営構想、中期経営計画の内容を考慮に入れた上で、モノ重視といった固定概念を超えた変革を目指し、制御事業にとどまらない将来に向けた全社レベルのブランドの基軸を明確にするためのコーポレートブランディングを進めていくという基本方針をまとめたのである。

nificantly. This reaffirmed the need for brand reestablishment.

2) What Are Yokogawa's Intentions as a Company?

Next, through interviews with management and an employee opinion survey, the team verified Yokogawa's direction, the management team's intentions, and employee views. An important part of this was gaining insight into whether, in addition to the management team, rank-and-file employees were aware of Yokogawa's reason for existing (namely, the company's aspirations and ideals), and whether this was being reflected in their conduct.

In particular, employees outside Japan voiced the desire for Yokogawa to achieve a more innovative and modern image, and were of the opinion that a new brand positioning was needed. These findings made clear the need for Yokogawa to move beyond the image of being a product manufacturer and become known as a solutions provider and a business partner to its customers.

3) What Core Competencies Should the Company Enhance to Differentiate Itself?

Based on the findings from 1) and 2), the team considered what the Yokogawa Group should do to strengthen its unique value proposition, while at the same time performing a competitive benchmarking analysis. From this, the team concluded that enhancing Yokogawa's strengths as a solutions provider and business partner, building up an image of innovation and leadership, and maintaining the association with high quality and strength in execution that it has cultivated with customers would help to give the company a long-term competitive advantage.

After taking into consideration the need for integration with the 100th Anniversary Project, the long-term business framework (discussed later), and the content of the mid-term business plan, the team put together a corporate branding roadmap to establish a foundation for the future of the brand at a company-wide level. This aimed for a transformation that went beyond the fixed notion of Yokogawa being solely focused on products, and had a wider focus than the industrial automation and control business.

経営層を巻き込んだコーポレートブランディングの推進
Corporate Branding Involving the Management Team

コーポレートブランディングでは、経営層を巻き込んだ全社的なブランディングを進めていくことになる。ブランド価値が事業部や製品に分散している場合には、その価値を一度集約し、全社レベルのブランドを確立した上で、各事業部や製品に落とし込んでいくという発想が必要なためである。その過程では、経営層による理解やサポートが不可欠だ。

横河電機では、2015年に向けて「長期経営構想」と「中期経営計画」の策定が進められていた。ブランディング推進チームは、経営層や策定チームが進めるそれらの活動と連携を図りつつ、本格的なブランディングに乗り出していく。

「長期経営構想」は、10年後のYOKOGAWAグループの「ありたい姿」について経営陣の考え方を定めたものであり、その中で「顧客と共に新しい価値を創造するソリューション提供企業を目指す」というメッセージが明確に示された。

そして、そのメッセージは、「YOKOGAWAは"Process Co-Innovation"を通じて、お客様と共に明日をひらく新しい価値を創造します。」というビジョンステートメントに昇華された。Process Co-Innovationとは、あらゆる情報やモノの流れを最適化して顧客と新しい価値を創造するソリューション全般を表した造語で、これが創立100周年を機に発表されるコーポレート・ブランド・スローガンにつながった。

Corporate branding activities were implemented company-wide and involved the management team. The reason for this is that, before one can disseminate and integrate brand value into business units and products, it first must be consolidated and established at a company-wide level. The understanding and support of the management team were essential in this process.

Looking toward 2015, Yokogawa was engaged in planning for its long-term business framework and mid-term business plan. The Branding Promotion Team linked its activities with those of the management team and the planning teams, and embarked on full-fledged branding.

The long-term business framework defines the management team's thinking on where the Yokogawa Group aims to be as a company ten years in the future, and clearly communicates that the company aims to be a solutions provider that creates new value together with its customers.

This framework's Vision statement reads, "Through Process Co-Innovation, Yokogawa creates new value with our clients for a brighter future." "Process Co-Innovation" refers in general to the creation of solutions with customers that generate new value by optimizing the flow of all kinds of information and goods. This vision ultimately led to the formulation of the corporate brand slogan that was announced on the occasion of Yokogawa's 100th anniversary.

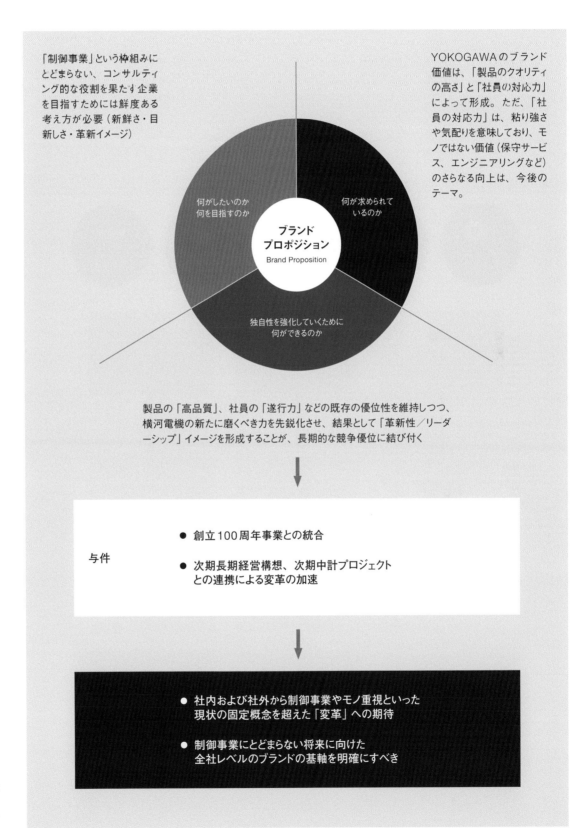

「制御事業」という枠組みに
とどまらない、コンサルティ
ング的な役割を果たす企業
を目指すためには鮮度ある
考え方が必要（新鮮さ・目
新しさ・革新イメージ）

YOKOGAWAのブランド
価値は、「製品のクオリティ
の高さ」と「社員の対応力」
によって形成。ただ、「社
員の対応力」は、粘り強さ
や気配りを意味しており、モ
ノではない価値（保守サービ
ス、エンジニアリングなど）
のさらなる向上は、今後の
テーマ。

何がしたいのか
何を目指すのか

何が求められて
いるのか

ブランド
プロポジション
Brand Proposition

独自性を強化していくために
何ができるのか

製品の「高品質」、社員の「遂行力」などの既存の優位性を維持しつつ、
横河電機の新たに磨くべき力を先鋭化させ、結果として「革新性／リーダ
ーシップ」イメージを形成することが、長期的な競争優位に結び付く

与件

● 創立100周年事業との統合

● 次期長期経営構想、次期中計プロジェクト
　との連携による変革の加速

● 社内および社外から制御事業やモノ重視といった
　現状の固定概念を超えた「変革」への期待

● 制御事業にとどまらない将来に向けた
　全社レベルのブランドの基軸を明確にすべき

ブランドプロポジション整理の検
討結果

Results of the brand proposition analysis review

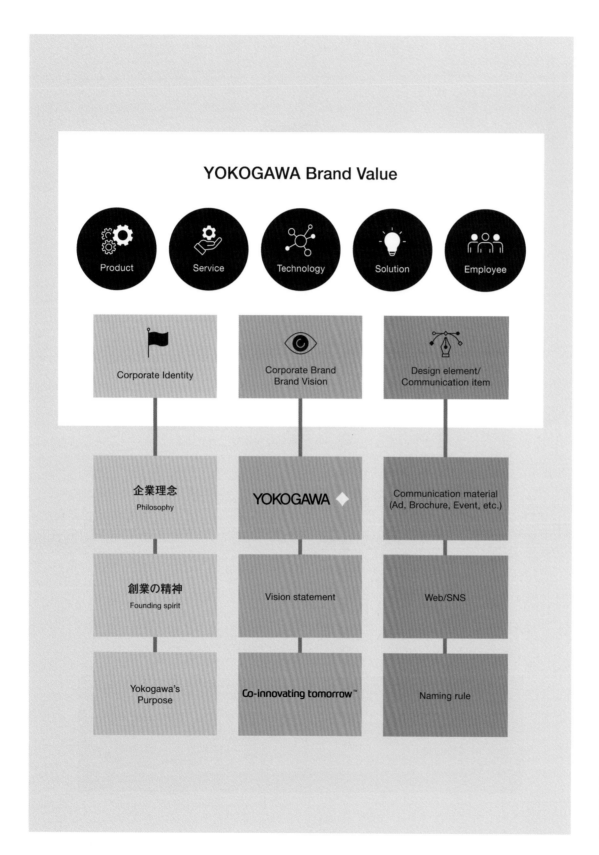

YOKOGAWA Brand Value

Product Service Technology Solution Employee

Corporate Identity	Corporate Brand Brand Vision	Design element/ Communication item
企業理念 Philosophy	YOKOGAWA ◆	Communication material (Ad, Brochure, Event, etc.)
創業の精神 Founding spirit	Vision statement	Web/SNS
Yokogawa's Purpose	Co-innovating tomorrow™	Naming rule

コーポレートブランドとしての
YOKOGAWAの構成要素の
整理

Review of Yokogawa's cor-
porate brand components

Contribution

中央大学 名誉教授

田中 洋 氏

日本企業に求められる"ブランド経営"
重要なのは継続性とチューニングのバランス

Profile

京都大学博士（経済学）。マーケティング論、ブランド論専攻。電通で21年間マーケティングディレクターとして実務を経験後、法政大学経営学部教授、コロンビア大学客員研究員、中央大学ビジネススクール教授を経て、2022年より現職。日本マーケティング学会会長、日本消費者行動研究学会会長を歴任。主著に『ブランド戦略論』（2017、有斐閣）がある。

企業にとって、なぜブランドが重要なのか。それは、ブランド価値が高いほど顧客に選択されやすくなるためである。さらに、ブランドが顧客の選択に使われるのは、顧客が購買前にブランドによって選択の判断をしなければならない場合である。例えば、製品の機能は事前にチェックできるが、アフターサービスなどの属性は、購買前のチェックが困難である。つまり、BtoBのような"理性的"な取引であっても、ブランドを選択の判断として使わなければならない場面があるということになる。ことに、取引相手に長期的な信用があるかどうかはブランドによる判断が大きい。近年は、BtoBの取引でも対面ではなくオンラインの活用が進んでいることから、信用の証であるブランドの重要性はさらに高まっている。

ただし、BtoC企業と比べてBtoB企業のブランディングには特有の難しさがある。BtoC企業では製品ごとのブランドマネージャーなどが決定権を持つことも多いが、コーポレートブランディングとなることが多いBtoB企業では、経営層が関与することになる。そのため、意思決定のプロセスが複雑になり、決定までの時間も長くなりがちだ。

コーポレートメッセージが"無難"なものになりがちな点も、BtoB企業のブランディングの難しさだ。コーポレートメッセージの決定では関わるメンバーが多くなり、その過程で当初の自社の特徴を表した案が、「信頼」「高品質」といった誰もが合意できる無個性なものになってしまいがちだ。

企業が発信するコーポレートメッセージで大事なことは、"現場で使えるかどうか"である。それは"耳に心地よい無難なもの"ではなく、現場のスタッフが困難に直面したとき、何を優先順位として判断するかを教えてくれるようなメッセージのことである。優先すべきものが明確であれば、現場の意思決定に反映でき迅速な問題解決が期待できる。

グローバルに展開する日本のBtoB企業を見ると、ブランド価値の形成に苦心している企業は多い。マーケティングが上手ではないため、海外市場から「付き合いにくい」と見られることも少なくない。必要なのは、グローバルに通用する"企業の顔づくり"だと私は考えている。

企業にも人格がある。発信するメッセージも企業やブランドの人格をイメージしてもらうことができれば、メッセージ単体よりも顧客の心に入っていける。グローバルな活動では、そうした"企業人格の形成"が大切になってくる。

横河電機は、早い時期からグローバル市場での競争において、ブランドの重要性を学んで、実践してきた数少ない日本企業の一つだと思う。印象的なのは、それが持続的であることだ。いくつかの企業では、よいメッセージを発し、よい活動をしても、経営層が変わると活動方針も変わってしまう例がある。ブランドのメッセージは継続することで"本物"になるのだ。

ただし、ブランディングには継続性だけでなく環境の変化に対応したチューニングも欠かせない。横河電機の取り組みでも、地球環境への意識の高まりを受け、「Co-innovating tomorrow」を使い続けながら「地球の物語の、つづきを話そう。」というキャンペーンも始めている。パーパスには「測る力」という横河電機としての原点が改めて入っている点にも注目したい。

今後の日本企業に望むのは、ブランド価値を生かした経営への取り組みである。ブランド価値の形成には時間がかかる。ブランドが必要になったときに活用できるよう、経営のスキルとして継続的にブランディングに取り組んでいただければと思う。

Contribution

Professor Emeritus, Chuo University

Hiroshi Tanaka

Brand Management: A Requirement for Japanese Companies
The Importance of Balance between Continuity and Fine-tuning

Why are brands important to companies? The primary reason is that the greater the brand value, the more likely the brand will be chosen by customers. Furthermore, brands help customers choose products by helping them make pre-purchase decisions. For example, product features can be checked in advance, but attributes like after-sales service are difficult to evaluate before a purchase. Even where transactions are "rational," as in BtoB contexts, brands must sometimes be included in the selection criteria. In particular, brands loom large in judging whether or not transaction counterparts can be trusted long-term. In recent years, with BtoB transactions conducted increasingly online rather than face to face, the importance of the brand in certifying trustworthiness has further increased.

However, compared to BtoC companies, BtoB branding involves special difficulties. In BtoC companies, individual product brand managers and others often have decision-making authority, but in BtoB companies, which often require corporate branding, upper management is often involved. The decision-making process thus tends to be more complex, and the time required to reach decisions longer.

Another difficulty in branding BtoB companies is that their corporate messaging often tends toward "safe." Numerous individuals become involved in shaping the company's messaging, and in the process, proposals for conveying the company's characteristics tend to start with attributes like "trustworthy," "high quality," and the like. Although these are attributes everyone can agree on, they are not distinctive and differentiating from their competitors.

What is important is that corporate messaging be usable on the front line. Rather than messages that are "easy on the ears," what is needed are messages that inform frontline staff how to set priorities when facing difficulties. Once priorities are clear, they can be reflected in frontline decision-making, and prompt problem-solving can be expected.

Many Japanese BtoB companies that are expanding internationally struggle to form brand value. Since marketing is not their strong suit, they are often seen as "difficult to deal with" by international markets. What is needed, in my view, is for these companies to create a corporate face that is effective globally.

Companies have personalities too. If a company's messaging enables customers to grasp its personality or that of its brand, the messaging will be more likely to stick in their minds compared to messaging in isolation. Such corporate personality formation becomes important when acting globally.

Yokogawa Electric Corporation is one of the few Japanese companies that learned early about the importance of branding and its uses in global competition. What is impressive is that this effort has been sustained. Some companies may have good messaging and related activities, but their policies change when management changes. Brand messaging becomes genuine with continuity.

However, branding requires not only continuity, but also fine-tuning to adjust to changes in the environment. In response to growing global environmental awareness, Yokogawa has launched a campaign called "What's next for our planet? Let's make it smarter." while maintaining the corporate brand slogan "Co-innovating tomorrow." I would also like to note how Yokogawa's Purpose affirms its foundation as a company with the "ability to measure."

Going forward, I hope Japanese companies will engage in management activities that leverage brand value. Forming such value takes time. I hope companies will work on branding continuously as a management skill, so they can leverage their brand when it becomes necessary.

Profile

Dr. Tanaka earned his Ph.D. in economics from Kyoto University, with specialties in marketing theory and brand theory. After 21 years of practical experience as a marketing director for Dentsu Inc., he went on to professorships at the Hosei University Faculty of Business Administration and the Chuo Graduate School of Strategic Management, and engaged in research as a Visiting Scholar at Columbia University before assuming his current position in 2022. He has served as chairman of the Japan Marketing Academy and of the Japan Association for Consumer Studies. His principal publications include Integrated Brand Strategy: Theory, Practice, & Cases (2017, Yuhikaku).

Chapter 2

初 動
The First Moves

「Co-innovating tomorrow」
の誕生

The Birth of "Co-innovating tomorrow"

コーポレート・ブランド・スローガンの発表
Announcement of the Corporate Brand Slogan

横河電機は2015年9月1日に創立100周年を迎えた。その記念すべき日に合わせて発表されたのが、コーポレート・ブランド・スローガン「Co-innovating tomorrow」である。

同年5月に発表された長期経営構想内で、前述のビジョンステートメントが提示されていた。これを受けて、ブランディング推進チームは、ビジョンステートメントを端的に表現するスローガンの策定に取り組んでいた。目的は、ビジョンステートメントをワンフレーズでシンボリックに表現し、YOKOGAWAグループ全社員がビジョンを共有できるようにすることだ。

こうして誕生したのが、"明日に向かって、様々な立場の人たちと対話しながら、新しい価値を「共」に「創」り上げ、イノベーションを共に巻き起こしていこう"という意味を持つCo-innovating tomorrowだ。「Co-innovating」には、お客様と共に長期的なパートナーシップを育みながら、課題解決のための新しい価値を共創すること、それによってステークホルダーと共にイノベーションを実現していくという強い思いが込められている。「tomorrow」は、着実に一歩一歩積み重ねていくことこそが、明日という未来に結びつくという信念を表している。未来を直訳すれば"future"だが、あえて"tomorrow"としたことは、信頼と堅実さを重んじる横河電機らしい選択だった。

ロゴのビジュアルにも、その思いと信念を込めた。Coは、Cとoが一体となることで顧客をはじめとするステークホルダーとのつながりを表現しており、その形が無限を意味する「∞」をメタファーとすることで、お客様との関係を永く継続していくという意志を表明している。

書体はPrometo(プロメト)をベースにした。「この書体は、角張っている正確性や精緻な印象を与える部分と、丸みを帯びた柔軟性という対照的な特徴を併せ持っています。この書体の選定には、グローバルを見据えつつローカルにも即してビジネスを進めるという両極にある課題の解決が、現代の企業の発展にとって重要であるという思いが込められています」とデザインを担当したメンバーは解説する。

そして、カラーロゴでは、"Co-innovating"と比較

Yokogawa Electric Corporation marked its 100th anniversary on September 1, 2015. The corporate brand slogan, "Co-innovating tomorrow," was announced on this occasion.

The Vision statement was released as part of the long-term business framework announced in May of the same year. Following this, the Branding Promotion Team worked on creating a slogan that succinctly summarized the ideas expressed in the Vision statement. The goal was to express the concept of the Vision statement in a single phrase that could be easily understood by all Yokogawa Group employees.

The result was "Co-innovating tomorrow," signifying the joint creation of new value and engagement in innovation while engaging with stakeholders and looking to the future.

"Co-innovating" conveys the determination to engage in the co-creation of value through the development of solutions in long-term partnerships with customers, and to realize innovation from this together with stakeholders. "Tomorrow" expresses the resolve to move steadily into the future one step at a time. While the word "future" could have been used, the word "tomorrow" was chosen intentionally as it was felt that it best embodied the trust and dependability that Yokogawa values.

The logo featuring this slogan also communicates this sentiment. The "c" and "o" in "Co" form a continuous element to express connection to customers and other stakeholders. The shape calls to mind "∞," the symbol for infinity, to express Yokogawa's intention to maintain relationships with customers in perpetuity.

The typeface is based on Prometo.

According to a team member who worked on the design, "This typeface is angular. It evokes accuracy and precision, but also has a roundness that signifies flexibility. It has both of these contrasting features. In selecting this typeface, we wanted to convey the importance to modern-day company development of resolving issues at both ends of the business spectrum; that is, of both executing business locally and maintaining a global focus."

In the logo, "tomorrow" is rendered in a lighter color than "Co-innovating," symbolically expressing

して "tomorrow" の文字の明度を上げたことで、明るい未来を象徴的に表現している。

この コーポレート・ブランド・スローガンは、社内外、国内外において強い支持と高い評価を受け、YOKOGAWAのブランドイメージ一新と認知度向上に大きな役割を果たしていくことになる。

a bright future.

This corporate brand slogan would go on to earn robust support and high marks both in- and outside the company as well as in- and outside Japan. It would also play a key role in updating Yokogawa's brand image and increase its brand recognition.

つながりを表現
Connect with clients/customers
Connect business process and business

あかるい未来
for a brighter future

Co-innovating tomorrow™

無限大 ∞

両利きスタイルの書体
cross-dominance
四角と丸の良さを兼ね備える

● "Co-innovating" には、お客様と共に長期的なパートナーシップを育みながら、課題解決のための新しい価値を共創していくというYOKOGAWAの強い思いが込められています。

● "tomorrow" は、着実に一歩一歩積み重ねていくことこそが、明日という未来に結びつくという信念を表しています。

コーポレート・ブランド・スローガンCo-innovating tomorrow
の要素分解

Elements of the corporate brand slogan, "Co-innovating tomorrow"

ブランドとアイデンティティのまとめ直し

Updating the Brand and Identity Together

横河電機には、創立から100年にわたる歴史の中で、大切に受け継いできた「創業の精神」「企業理念」「共有する価値観」があった。こうした既存の体系と、ブランド再構築の過程で策定した「ビジョンステートメント」や「コーポレート・ブランド・スローガン」の関係性を社員に理解してもらうため、しっかり説明する必要性を感じていた。

そこでブランディング推進チームは、ブランドとアイデンティティを体系的に整理し、「Yokogawa's Brand & Identity」としてまとめ直す作業を進め、創立100周年という機会で発表した。

Yokogawa's Brandは、会社を表し重要なコミュニケーション上のシンボルとしてのコーポレートブランド、コーポレートブランドであるYOKOGAWAのブランド価値を向上させるため、様々なステークホルダーとの接点で一貫して伝えていくコーポレート・ブランド・スローガン、および将来に向けたありたい姿、企業としての理想を端的に示すビジョンステートメントの3つにより構成されている。

Yokogawa's Identityは、時代が変わっても継承していくべきアイデンティティとして、「創業の精神」「企業理念」「共有する価値観」の3つを掲げた。これらは優れたチーム力を発揮するために、グループ社員全員が行動する際のよりどころである。

Over the more than 100 years since Yokogawa's founding, the company has stayed true to its Founding Principles, the Yokogawa Philosophy, and its Core Values. However, it was decided that extensive briefings were needed to ensure that employees understood the relationship between these existing schemes, the Vision statement, and the corporate brand slogan that were created in the brand reestablishment process.

The Branding Promotion Team then systematically structured the brand and identity, and updated them together as "Yokogawa's Brand & Identity," which were unveiled on the occasion of the 100th anniversary.

Yokogawa's Brand consists of three elements: 1) the corporate brand as a symbol used in important company communications that expresses the company, 2) the corporate brand slogan that is consistently communicated to various stakeholders to increase the value of the Yokogawa corporate brand, and 3) the Vision statement that succinctly expresses the company's ideals and aspirations for the future.

Yokogawa's Identity has been preserved through changing times and encompasses the three elements of the Founding Principles, the Yokogawa Philosophy, and the Core Values. These form the cornerstone of Group employees' conduct and empower them to demonstrate exceptional teamwork.

Yokogawa's Brand & Identity

創立以来100年にわたる歴史の中で、
大切に受け継いできた創業の精神、そして、
企業理念、コーポレートブランド（トレードマーク）と、
これからのあるべき姿を示したビジョンやスローガン。
YOKOGAWAグループを創りあげる要素を
「Yokogawa's Brand & Identity」として
改めて整理しました。

Yokogawa's Brand

コーポレートブランド
Corporate brand

コーポレート・ブランド・スローガン
Corporate brand slogan

ビジョンステートメント
Vision statement

Yokogawa's Identity

創業の精神
Founding principles

企業理念
The Yokogawa Philosophy

共有する価値観
Core values

Yokogawa's Brand

お客様に対する責任の表れであるとともに、
お客様やパートナー様からの信頼と
満足の証しともなるYOKOGAWAのブランド。
グループ社員はブランドに込められた思いを理解し、
日々の活動においてお客様、パートナー様との絆、
およびYOKOGAWAブランドを
育てていきます。

Yokogawa's Brand

コーポレートブランド
Corporate brand

YOKOGAWA ◆

コーポレート・ブランド・スローガン
Corporate brand slogan

Co-innovating tomorrow™

ビジョンステートメント
Vision statement

YOKOGAWAは "Process Co-Innovation" を通じて、
お客様と共に明日をひらく新しい価値を創造します。

Through "Process Co-Innovation,"
Yokogawa creates new value with our clients for a brighter future.

Yokogawa's Identity

お客様の夢に貢献したい、
豊かな社会づくりに尽くしたい。
時代が変わっても、これからも変わることなく
継承していくべきYOKOGAWAのアイデンティティ。
地域や事業分野を超えて、
優れたチーム力を発揮するために、
グループ社員全員が行動する際の拠りどころです。

Yokogawa's Identity

創業の精神
Founding principles

・品質第一主義・パイオニア精神・社会への貢献

企業理念
The Yokogawa Philosophy

YOKOGAWAは計測と制御と情報をテーマに
より豊かな人間社会の実現に貢献する

YOKOGAWA人は良き市民であり
勇気をもった開拓者であれ

共有する価値観
Core values

・個の尊重・価値共創・コラボレーション・誠実・感謝のこころ

創立100周年事業における情報発信

Communicating Information in the 100th Anniversary Project

創立100周年事業は、横河電機単体の100周年ではなく、国内外グループ会社も巻き込んで世界中のお客様へ感謝を伝える、グループ全体の100周年と位置づけられた。ブランディング推進チームが発信するビデオや広告なども、グローバルでの使用を大前提にして制作が進められた。

2015年5月に公開した創立100周年をテーマとしたビデオは、従来のメーカーのイメージから脱却し、新たな挑戦をするYOKOGAWAグループの姿を強く意識している。

このビデオは「Tomorrow's in sight.」というフレーズで締めくくられており、ここに二重の意味を込めている。一つは「明日に向けた私たちの知見、洞察」ということ、もう一つは「明日という未来は、私たちの視野にすでに入っている」である。これが、9月に発表されるコーポレート・ブランド・スローガンCo-innovating tomorrowのプロローグとして使用された。

100周年の当日にあたる2015年9月1日付の新聞各紙には、Co-innovating tomorrowを初めて対外的に打ち出した広告を掲載。これにより次の100年に向かう新しい横河電機を知ってもらうとともに、新しいコーポレート・ブランド・スローガンを広く認知してもらうという目的があった。

広告は日本人と外国人の2人の子どもが力を合わせ、積み木でCo-innovating tomorrowを形作るビジュアルによって、世界中の顧客と信頼関係を築きながら、新たな価値を共創している姿を表現している。子どもをモデルとすることで自由な発想を、文字の素材に積み木を使うことで一歩一歩の積み重ねをシンボライズした。

そして、100周年記念事業の最後を飾ったのは、お客様をお招きして11月に開催した式典であった。2015年におけるブランド再構築の活動は、100周年記念事業とリンクしながら着実に歩みを進めていったのである。

The 100th Anniversary was positioned as a celebratory occasion for the entire Yokogawa Group, not just Yokogawa Electric, and as an opportunity to involve Group companies in- and outside Japan in communicating their gratitude to customers around the world. Videos and ads released by the Branding Promotion Team were created on the assumption that they would be used globally.

A video about Yokogawa's 100th anniversary that was released in May 2015 focused on transforming the Yokogawa Group's image from traditional manufacturer to an organization that seeks out new challenges.

The video ends with the line "Tomorrow's in sight," which has two meanings. The first is that Yokogawa has the knowledge and foresight to visualize tomorrow, and the second is that Yokogawa already has the future in its sights. This line served as a prologue to the "Co-innovating tomorrow" corporate brand slogan that was unveiled in September 2015.

On September 1, 2015, the date of Yokogawa's 100th anniversary, the company placed newspaper ads that for the first time publicly used the "Co-innovating tomorrow" slogan. The goal was to introduce the public to a new Yokogawa that was embarking on its next 100 years and to gain widespread recognition of the new corporate brand slogan.

The ad showed two children, one Japanese and one from another country, working together with blocks to create a visual in the shape of "Co-innovating tomorrow" that expressed Yokogawa's aspiration to create new value while building relationships based on trust with customers all around the world. Children were used in the ad to symbolize creative thinking, while the blocks symbolized building things step by step.

The final jewel in the crown of the 100th Anniversary Project was a November ceremony to which customers were invited. The brand reestablishment activities in 2015 were carried out systematically in conjunction with the 100th Anniversary Project.

一緒なら、どんな未来も創っていける。

Co-
innovating
tomorrow

Co-innovating tomorrow
このコーポレート・ブランド・スローガンは、
次代をひらくイノベーションをお客様とともに、という
私たちの次の100年に向けた約束です。
この約束を果たしていくために、私たちは、
計測・制御・情報を組み合わせたソリューションで、
業界や国境を超えて「モノ」をつなぎ、
これまでにない「コト」や価値を創り出していきます。
こうした挑戦の積み重ねによって、
明日という未来はより確かなものになるでしょう。
世界中のお客様と、「夢」をかなえる歩みをともに。
イノベーションのパートナーはYOKOGAWAです。

Co-innovating tomorrow™

YOKOGAWA | 100th ANNIVERSARY 1915-2015

おかげさまで、横河電機株式会社は
本日創立100周年を迎えることができました。

www.yokogawa.co.jp 横河電機株式会社

Tomorrow's in sight.

YOKOGAWA | 100th ANNIVERSARY 1915-2015

従来のメーカーのイメージか
ら脱却し、新たな挑戦をする
YOKOGAWAグループの姿
を訴求した創立100周年ビデオ
（2015年5月公開）

A video made for the
100th anniversary promot-
ing Yokogawa Group's
transformation from a traditional manufacturer to
an organization that seeks
out new challenges (re-
leased in May 2015)

2015年11月に開催した創立
100周年イベントの様子と創立
100周年記念ロゴを使用した
各種制作物

An event for the 100th anniversary held in November
2015 and products featuring
the 100th anniversary logo

キー・デザイン・エレメント「Brilliant Grid」の誕生
The Birth of the Key Design Element, the Brilliant Grid

100周年記念事業が一段落したところで、ブランディング推進チームが取り組んだのは、キー・デザイン・エレメントの制作である。YOKOGAWAを強いブランドにするため、コーポレート・ブランド・スローガンをビジュアルで訴求し、企業全体で統一したイメージを創出することが目的だ。Co-innovating tomorrowは、ロゴも合わせて発表をしてはいたが、それが意味する価値共創、未来志向をイメージするキー・デザイン・エレメントが求められていた。それは、制御事業や製品ブランドの上位に位置し、コーポレートブランド全体を包含するものである。そのため、デザインの選定にあたっては社内外で活発な意見交換を重ね、グローバルで戦えるイメージを目指した。

こうして2016年4月に発表されたのが、Brilliant Grid（ブリリアントグリッド）である。

YOKOGAWAを象徴するシンボルカラーの黄色い正方形を中心に据え、光が放射するイメージをビジュアル化した白色の透明感あるグラデーションで、周囲に複数の正方形を配置したデザインだ。

正方形が規則正しく重なる様子は、顧客から評価されている安定性や技術の正確性とともに、サステナビリティを表現している。光の放射は輝く明日やビジネスの変革と革新の広がりを表し、広がるにつれて透過性を持ち背景の画像や色と重なり合って一つになる姿は、顧客との共創を意味したものである。

As the 100th Anniversary Project drew to a close, the Branding Promotion Team's next project was to create a key design element. The goal was to produce a single corporate image that visually aroused interest in the corporate brand slogan and created a strong impression of the Yokogawa brand. Yokogawa already had announced the logo for "Co-innovating tomorrow," but a new key design element that could be associated with Yokogawa's value co-creation and future-oriented aspirations was also needed to go along with that. This element would be positioned above the industrial automation and control business and product brands, and encompass the corporate brand as a whole. As part of the design selection process, a series of stimulating discussions were held in- and outside the company to exchange ideas on how to create an image of global competitiveness.

Out of this process came the Brilliant Grid, which was announced in April 2016.

The design uses a central square rendered in yellow, Yokogawa's color. Surrounding the square are a number of other squares that gradually become more transparent and whiter to visually depict light radiating from the yellow square.

The regularity of the placement of the squares expresses the stability and technical precision that Yokogawa is known for as well as its sustainability as a business. The radiating light expresses a bright tomorrow as well as expanding business transformation and innovation. Superimposing the design over background images and colors as it gradually becomes more transparent and having the design layers blend together signifies co-creation with customers.

Brilliant Grid の発表に合わ
せて公開した「横河グループコ
ミュニケーション・デザイン・ガ
イドライン」と、その中に記載さ
れたBrilliant Gridに込めた想
い

Yokogawa Group Communi-
cation Design Guidelines released with the unveiling of
the Brilliant Grid, and the in-
tent behind the Brilliant Grid
included in the guidelines

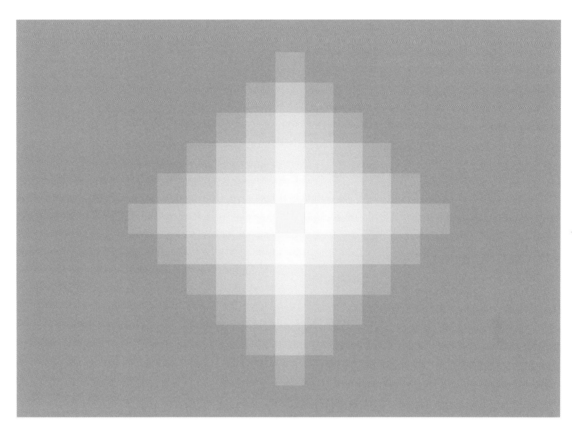

ブリリアントグリッド

ブリリアントグリッドはYokogawa yellowの正
方形（イエロースクエア）の中心から放射状に
広がる光（ホワイトスクエア）のイメージをビジュ
アル化したキー・デザイン・エレメントです。

Shine（明日をひらく）

イエロースクエアの中心から光を感じさせるホワイトスクエア
のグラデーションから成り、輝く明日という未来を表現すると
共に、サスティナビリティ、技術の安定性、正確性を表し
ています。

Expand（変化する）

中心から一つひとつのスクエアが拡張していくグラフィック
は、YOKOGAWAのビジネスの変革と革新の着実な広が
りを表しています。

Synchronize（パートナー感）

ブリリアントグリッドが背景の色面またはフォトイメージの色
味の影響を受け重なり合いひとつになる姿は、お客様との
共創・パートナー感を表現しています。

言うまでもなく、使用する色彩や形を厳格に指定し、同時に使用する写真のトーン・アンド・マナー、使用フォントも規定。カタログやWebサイトへ展開する際の世界観などもガイドラインとして取り決めていった。

このころのブランディング推進チームにとっての課題は、グローバルへの展開だった。2016年4月1日に日本語版を公開するだけでも大変な作業だったが、それと同じものを英語でも制作しなければならなかったからだ。しかし、そんな危惧を打ち払ったのは、GMT（Global Marketing communication Team）の熱意であった。

「3月にGMT担当者とのミーティングがあり、その席でデザインガイドラインを紹介したところ、こちらが驚くほど好意的に受け入れてくれたのです。ドラフトを自国に持ち帰ってコミュニケーションマテリアルを展開する準備を進めてくれました。おかげで、海外でも日本と同時にBrilliant Gridを展開することができたのです」とメンバーの一人は語る。

日本では、コーポレートマーケティングや事業部マーケティングのコミュニケーションマテリアル制作やイベント運営のサポートを依頼しているパートナーにも、新しいガイドラインの説明会を実施した。実際に制作を担当する現場にガイドラインの内容を理解してもらうことで、新しいデザインエレメントへスムーズに移行でき、お客様との接点でのイメージの統一が進むと考えたためだ。

It goes without saying that the colors and shapes were selected with utmost care. At the same time, the tone and manner of the photos and font were also stipulated. The worldview and other elements to be expressed with the use of the Brilliant Grid in catalogs and websites were also specified in the guidelines.

A challenge the Branding Promotion Team faced around this time came with the global deployment of these materials. The release of the Japanese version on April 1, 2016 was a major undertaking, but the team also had to create the same materials in English. This was a daunting task, but the enthusiasm of the Global Marketing Communication Team (GMT) swept away the Branding Promotion Team's apprehensions.

"In March, we had a meeting with people from the GMT. When we presented the design guidelines to them, we were surprised at how well they were received. They then took the draft guidelines back to their own countries and used them to prepare communication materials. Thanks to this, we were able to deploy the Brilliant Grid globally at the same time it was deployed in Japan," explained one member of the Branding Promotion Team.

In Japan, briefings on the new guidelines were also held for partners who were asked to create communication materials for corporate marketing and business unit marketing activities and to support event operations. These were held with the belief that if the teams in charge of actually producing the materials understood the content of the guidelines, it would enable a smoother transition to the new design element and result in a more unified image at customer points of contact.

将来の社員もエンゲージしたブランディング

Branding that Also Engages Future Employees

Brilliant Gridを使用した最初の制作物は、2016年度の横河電機の入社案内（リクルートメント・ガイドブック）だった。前年度までは、いかにも計測機器・制御システムのメーカーらしい堅実な入社案内だったが、ソリューション提供企業としての成長、新事業への展開を目指す横河電機とのイメージの乖離が存在しつつあった。

新しい入社案内の表紙にはBrilliant Gridを描き、「心躍るイノベーションを、ともに」というコピーを配した。「横河電機が今後目指す方向が伝わる入社案内をつくりたい」というのが当時の採用担当者からのリクエストであった。

The first production that used the Brilliant Grid was Yokogawa's fiscal year 2016 recruitment guidebook. Up to the previous fiscal year, recruitment guidebooks conveyed the solidly dependable image expected of a manufacturer of measurement devices and control systems, but this was increasingly at odds with Yokogawa's targeted image of growing as a solutions provider and developing new businesses.

The Brilliant Grid was depicted on the cover of the new recruitment guidebook with the tagline, "Exciting innovation, together." The hiring manager at the time had requested that the recruitment guidebook convey Yokogawa's future direction.

Brilliant Gridが初めて使用された2016年度の入社案内

Recruitment guidebook from fiscal year 2016, when the Brilliant Grid was first used

もちろん、そこには世界を舞台にする横河電機で通用する人財を集めたいという思いがあり、いわば将来の社員をも巻き込んだブランド戦略であった。こうした入社案内のイメージ刷新に加え、Webサイトのリニューアルや人材部門による取り組みも相まって、女性の志望者や技術系以外の志望者が増えていく。

次に、社外に公開しているWebサイトもリニューアルした。100周年を機に2015年4月には小規模なリニューアルを行っていたものの、Brilliant GridとCo-innovating tomorrowのロゴをあしらい、顧客とのあらゆるタッチポイントにおいて統一したイメージを構築し、ソリューション提供企業らしくイメージを一新して2016年4月に公開した。

Yokogawa was doing business globally and wanted to attract world-class talent, so in a sense this was also part of its brand strategy involving future employees. The updated image in the recruitment guidebook, together with the redesign of Yokogawa's website and initiatives conducted by the HR department, proved more appealing to women and individuals from non-technical backgrounds.

Next, the external-facing website was also redesigned. While minor website updates were made in April 2015 for the 100th anniversary, the major redesign in April 2016 featured the Brilliant Grid and the "Co-innovating tomorrow" logo, built a unified image at all customer touchpoints, and created an image consistent with a solutions provider.

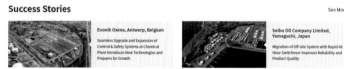

Brilliant Gridが初めて使用されたWebサイト（2016年4月公開）

Website using the Brilliant Grid for the first time (Released in April 2016)

Chapter 3

浸透
Penetration

社外への認知度向上と
社内への啓蒙活動

Enhancement of External Recognition and
Promotion of Internal Awareness

ブランド浸透のための様々な活動
Activities to Enhance Brand Recognition

2015年の創立100周年を機に、横河電機はコーポレート・ブランド・スローガン「Co-innovating tomorrow」の発表とブランドとアイデンティティの体系化を行った。ブランディング推進チームは、翌2016年に、横河電機の価値共創、未来志向への思いを伝えるキー・デザイン・エレメントの策定を行ったことを受け、ブランドのさらなる社外での認知度向上と社内での啓蒙活動に取り組んだ。

Co-innovating tomorrowに対する理解を深めてもらうための、キー・デザイン・エレメントであるBrilliant Gridを活用した広告やポスターの展開もそうした活動の一つである。

例えば広告では、Co-innovating tomorrowのロゴを「一歩ずつ、一緒に。一歩ずつ、未来へ。」というコピーと共に配置。ポスターではCo-innovating tomorrowに込めた想いを伝えるため、「お客様を想起させる産業プラント」「未来の象徴である子ども」「お客様とのコラボレーション」を題材に3種類の画像を選

In 2015, the year of its 100th anniversary, Yokogawa Electric Corporation announced its "Co-innovating tomorrow" corporate brand slogan and systematized its brand and identity. The following year, in response to the formulation of the key design elements that convey Yokogawa's value co-creation and future-oriented aspirations, the Branding Promotion Team worked to further increase external recognition and raise internal awareness of the brand.

One such activity was the deployment of advertisements and posters utilizing the Brilliant Grid, a key design element, to deepen understanding of "Co-innovating tomorrow."

For example, in advertisements, the "Co-innovating tomorrow" logo was paired with the ad copy "Step by step, together. Step by step, into the future." For posters, three images were selected to convey the idea behind "Co-innovating tomorrow," based on the themes of "Manufacturing plants that call to mind our customers," "Children are a symbol of the future," and "Collaboration with customers."

「産業プラント」「未来ある子ども」「コラボレーション」をテーマとしたポスター

Posters with the themes of "Manufacturing Plant," "Children with a Future," and "Collaboration"

定した。

「Co-innovating tomorrow や Brilliant Grid の発表に合わせ、ルールやテンプレートをまとめ、『コミュニケーション・デザイン・ガイドライン』として公開しました。どのような文化や習慣を背景としていても、世界各地で統一したブランドルールに基づいた制作物を作成できるようになりましたが、まずは私たちがこのガイドラインに沿った、グローバルでの活用を大前提に新しいクリエイティブを制作し、それを各国でも利用してもらえるようにしたのです」

　Brilliant Gridは、印刷物だけではなく立体的な表現にも適用された。例えば、2017年に日本で開催された大規模展示会に参加した際には、会場内の横河電機のブースにBrilliant Gridを立体的に表現した四角や格子状のデザインの椅子や調度を配置した。プレゼンテーションのテンプレート、印刷物、映像などを統

"In conjunction with the launch of 'Co-innovating tomorrow' and the Brilliant Grid, we have compiled a set of rules and templates and published them as the Communication Design Guidelines. We can now create communication materials globally, based on uniform branding rules, regardless of local cultures and customs. But first, we issued new creatives in accordance with these guidelines, that can be used in any country, with a basic premise of global use."

The Brilliant Grid was applied not only to printed materials, but also to physical items that were on display. For example, at a major Japan exhibition in 2017, the company's booth featured chairs and other furnishings with square and lattice patterns based on the Brilliant Grid. Presentation templates, printed materials, videos, and other communications tools were all developed with a unified image, to enhance brand appeal.

With respect to boosting external awareness, board members—who are an important point of

2017年に日本で開催された展示会に参加した際にはBrilliant Gridを空間デザインにも使用

The Brilliant Grid was also used in the spatial design for Yokogawa's participation in an exhibition held in Japan in 2017

子供たちの笑顔

この星のために

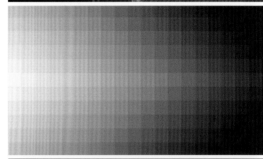

「Co-innovating tomorrow」
TVCF 子供たちの笑顔篇
Brilliant Gridから企業ロゴま
でのモーショングラフィックスを
使用

"Co-innovating tomorrow"
TVCF - Children's smiles
Motion graphics from Brilliant Grid to corporate logo

未来世代と分かち合えるよう

YOKOGAWA ◆
Co-innovating tomorrow™

社員にステークホルダーとの共
創を意識してもらえるように、本
社のエレベーターや壁面などを
Co-innovating tomorrow
のロゴとBrilliant Gridでラッピ
ング

Wrapping the elevators and walls of the headquarters with the "Co-innovating tomorrow" logo and Brilliant Grid to strengthen employee awareness of co-creation with stakeholders

一したイメージで展開することで、ブランド訴求を高めた。

　社外への認知度向上という点では、顧客や投資家などのステークホルダー、報道機関との接点となる役員が果たす役割も大きい。そこで役員向けに「Yokogawa's Message House」という小冊子を日本語と英語の2言語で作成して配布した。ここには、横河電機のアイデンティティやブランドに対する考え方がまとめられており、社外に発信するメッセージの統一につながった。また、スピーチや取材対応などでCo-innovating tomorrowを繰り返し伝えるよう依頼しブランド浸透も図った。

contact with stakeholders such as customers and investors as well as with the press—play a key role. To this end, the company prepared and distributed Japanese and English versions of a "Yokogawa Message House" booklet to its executives. It profiles Yokogawa's identity and brand approach, and promotes consistent external messaging. In addition, board members are encouraged to use our "Co-innovating tomorrow" corporate brand slogan in speeches, interviews, and other outward-facing contexts to drive brand awareness.

役員向けに横河電機のアイデンティティやブランドに対する考え方をまとめた小冊子「Yokogawa's Message House」

"Yokogawa's Message House," a booklet for executives that profiles Yokogawa's identity and branding approach

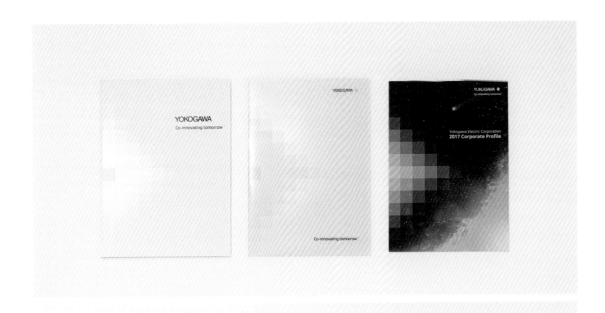

Co-innovating tomorrow
と Brilliant Grid を使用した
コーポレート、事業部コンセプ
ト、製品/ サービスの各レイヤー
のカタログ

Catalogs for the company,
business unit concepts, and
products and services layers, using "Co-innovating
tomorrow" and the Brilliant
Grid

社員こそが「ブランディングの主人公」
Employees Are Brand Heroes

一般にブランディングというと、社外に向けた宣伝・広報活動によって企業や製品のイメージを高めることを連想しがちだが、それはブランディングの本質の一部でしかない。ブランディングを成功に導くには、社外向けのアウターブランディングだけでなく、社内向けのインターナルブランディングが不可欠であり、この2つを車の両輪のように進めていく必要がある。

横河電機のブランドの再構築にあたり、企業文化を変革する装置としてブランディングを活用しようとしていたブランディング推進チームがとりわけ重要視したのが、このインターナルブランディングである。社員が参加意識をもって積極的に関与するかどうかが、成否を握っていたと言っても過言ではない。

加えて、ブランディングを一過性のものに終わらせるのではなく、時代や市場の変化に合わせて刷新しながら継続的に進めることが、企業文化を変革する装置に求められる。そのための仕組みづくりが重要であると考えた。

そうした仕組みの一つとして、社内向けの「YOKOGAWAブランドサイト」による情報発信を進めた。本サイトの目的は、世界中の全社員とYOKOGAWAブランドに関する情報や考えを共有することにあった。そこには、ブランドとは何かについての基本知識の説明から始まり、YOKOGAWAのビジョンやスローガンを整理したYokogawa's Brand & Identityの解説が掲載された。

コーポレート・ブランド・スローガンCo-innovating tomorrowの社内発表は、社外に先駆けて行われた。その理由を担当者は次のように振り返る。
「私たちは、社員こそがブランディングの主人公であり、最も重要なブランドアンバサダーであると考えています。社外発表に先立ち、社員にコーポレート・ブランド・スローガンを提示したのもそのためです。このようにして、YOKOGAWAのブランドの方向性をいち早く知ってもらい、ブランディングを自分ごととして捉えてもらいたかったのです」

つまり、社員がブランディングに対する必要性や効果

In general, branding is associated with advertising and public relations activities that serve to enhance the image of a product or company. However, such outward-facing activities represent only one aspect of branding. To achieve branding success, internal branding is just as important. These approaches are like the two sides of a coin, and both must be pursued.

In its efforts to reestablish the corporate brand, the Branding Promotion Team gave a special emphasis to internal branding as it saw this as a means for the transformation of Yokogawa's corporate culture. It is not an exaggeration to say that the success or failure of the brand reestablishment effort turned on employee participation and active involvement in branding.

Beyond this, it is important to note that, as times change and markets evolve, branding is a means for the continual transformation and renewal of corporate culture, and not just some transient or temporary phenomenon. The team therefore believed it was important to create a system to sustain this effort.

One key element of this system was the Yokogawa brand site on the company intranet to disseminate information. The aim was to enable Yokogawa employees around the world to share brand-related information and opinions. The site offers basic information on what a brand is and explains Yokogawa's Brand & Identity, and also covers the company's Vision statement and corporate brand slogan.

The internal announcement of the "Co-innovating tomorrow" corporate brand slogan took place before the slogan was publicly announced. Here, the individual responsible for the announcement reminisces on the reason for this approach:

"We see our employees as branding heroes. They are our most important brand ambassadors. That is why we presented the corporate brand slogan to our employees before making it public. We wanted to ensure that employees were up to speed on the direction of the brand, and experienced branding as something with personal relevance."

について実感を持てるようにするために、YOKOGAWAブランドサイトを通じて会社としての本気度を示したのである。

その後、YOKOGAWAブランドサイトは、プレゼンテーションツールのテンプレートやグラフィックスの使用ガイドラインなどが追加・更新され、YOKOGAWAブランドの意味、背景にある考え、市場や社会へのコミットメントを社員と共有する機能を担うベースになっていくこととなる。

As such, the Yokogawa brand site demonstrates the company's commitment to branding and gives employees a first-hand sense of branding's necessity and effectiveness.

Going forward, the Yokogawa brand site will be augmented and updated with presentation tool templates and guidelines for the use of graphics. It will serve as a place where employees can learn about the meaning of the Yokogawa brand, the concepts behind it, and the commitment that it expresses to the market and society.

世界中の社員とYOKOGAWAブランドに関する情報や考えを共有するために開設した社内向けのYOKOGAWAブランドサイト

The Yokogawa's Brand Site is an intranet site created to enable employees worldwide to share information and ideas relating to the company brand

社員への啓蒙活動の強化
Strengthening Internal Activities to Boost Awareness

ブランディング推進チームが社員に理解してほしかったことの一つが、「ブランディングは、一過性のキャンペーンではなく継続する活動である」ということ。社員がブランドアンバサダーであるという点は、ブランディングの始動時からの基本姿勢であるが、それをさらに強化して、社員それぞれがブランディングを "自分ごと" とする方法を模索していった。

その一つが、社員に配布する「ブランド・アンバサダー・カード」の制作だ。ブランドを意識させるデザインのカードを名刺の箱1つに1枚ずつ封入した。アンバサダーカードは複数のデザインを用意し、コレクションを楽しみながら自然とブランディングを "自分ごと" として捉えてもらうことを狙った。

ブランドの象徴としての本社では、オフィスブランディングを施した。具体的には、日常的に社員にCo-innovating tomorrowを通じてステークホルダーとの共創を意識してもらえるよう、エレベーターや建物の入り口をラッピングして、ロゴと共にBrilliant Gridを掲示した。

2015年9月にイントラネットで公開したYOKOGAWAブランドサイトも、精力的にコンテンツの拡充・整備を行った。

Co-innovating tomorrowのロゴデータやブランディング推進チームが作成した動画データ、ポスターの画像、スクリーンセーバー用の素材などを格納していった。また、前述のコミュニケーション・デザイン・ガイドラインも、同サイトに掲載し、各事業部のマーケティングメンバーが率先して、新しいブローシャーに刷新を進めた。このガイドラインの「はじめに」として、以下のようなメッセージが記されている。ここからも、社員にブランディングを "自分ごと" として継続的に取り組んでほしいという思いが伝わる。

お客様は私たちの様々な活動を通じた体験からYOKOGAWAに対するブランドイメージを形成しています。製品、サービス、WEB、イベント、記事、カタログ、広告、社員の言動などの各種コンタクトポイントにおいて、統一感がとれていなかったり、短期的な視点で異

One thing that the Branding Promotion Team wanted employees to understand was that branding is not a one-off campaign, but rather is ongoing. While employees have been seen as brand ambassadors since the launch of the branding, the team sought ways to strengthen this understanding and give employees a sense that branding is relevant to them personally.

One project has involved the creation of Brand Ambassador Cards for distribution to employees. To raise brand awareness, each box of employee business cards contains a specially designed Brand Ambassador Card. There are multiple card designs, and employees are encouraged to collect them. In so doing, the hope is that they will perceive branding as something that is personally relevant.

As a symbol of the brand, the Yokogawa headquarters buildings were given a branding makeover. Specifically, to reinforce employee awareness of the concept of co-creation with stakeholders that is expressed with the "Co-innovating tomorrow" corporate brand slogan, elevator doors and doors at the main entrance were decorated with the Brilliant Grid and the brand logo.

Content on the Yokogawa brand site, which debuted on the intranet in September 2015, was also actively expanded and enhanced.

The "Co-innovating tomorrow" logo as well as videos, posters, and screen savers created by the Branding Promotion Team are available for download there. There is also a link to the aforementioned Communication Design Guidelines, and marketing personnel in each business unit have taken the initiative to upload their new brochures to the site. The following message is included in the introduction to the guidelines. This communicates the hope that employees will see branding as something that is personally relevant and remain engaged in this activity:

Building customer recognition of the Yokogawa brand depends on many different activities. Managing all the touchpoints where we encounter our stakeholders — products, services, web sites, events, news stories, pamphlets, advertisements,

なるコミュニケーションが行われると、お客様に
YOKOGAWAを十分理解いただくことはできません。

　このコミュニケーション・デザイン・ガイドラインは、異なるコンタクトポイントにおいても一貫性と継続性のあるコミュニケーションを展開するための指針を示しています。どのコンタクトポイントに触れても、お客様がYOKOGAWAに対し同じイメージを抱いてくだされば、YOKOGAWAブランドの存在感はより明確に、より強くなります。そして、YOKOGAWAブランドは、何よりも私たちにとって大きな価値となります。

　コミュニケーション・デザイン・ガイドラインに従って、適切なブランドコミュニケーションを推進してください。

　「YOKOGAWAブランドサイトは2023年現在も更新を続けています。社員は、Webページやプレゼンテーションの作成時にここにアクセスして確認するなど、ブランドに対する考え方や行動に変化が表れています。ブランディングが"自分ごと"になってきたと言えるでしょう」とメンバーの一人は語る。

and our employees themselves — is extremely important in implementing our brand strategy. If we present our brand inconsistently across these touchpoints, or communicate only with a short-term view, we will not convey a sense of what makes Yokogawa unique. When our stakeholders are able to see a unified Yokogawa brand, no matter which touchpoint they encounter, our presence will become clearer and stronger.

　These Communication Design Guidelines provide help on developing communications with consistency and continuity across all of these diverse touchpoints. Simply put, nothing is more valuable to us than the Yokogawa brand.

　Please use these Communication Design Guidelines to ensure that we always deliver appropriate communications that consistently differentiate the Yokogawa brand.

"The Yokogawa brand site continues to be updated as of 2023. Employees access the site to confirm adherence to policy when they create web pages and presentations, and changes in brand-related attitudes and actions are evident." As one team member from the Branding Promotion Team notes, "Branding has become personally relevant."

複数のデザインを用意し、名刺の箱に封入して配布した「ブランド・アンバサダー・カード」
Brand Ambassador Cards were created with multiple designs, and distributed individually in boxes of employee business cards

ブランド・キャンペーン・サイトの公開
Brand Campaign Website Launch

　社内向けのYOKOGAWAブランドサイトとは別に、現在の顧客だけでなく、将来の顧客にも、次の100年に向かう横河電機を理解し、新たなイメージを確立してもらうため、2016年9月、ブランド・キャンペーン・サイトを公開した。

　競合他社や他のグローバルBtoB企業がどのような内容のコンテンツをどのような見せ方でデジタルの世界で展開しているかを研究した上で、横河電機がお客様の課題を共に解決していく企業として、信頼できるパートナーであると認識してもらうことを目的とした。そのため、内容は製品カタログのような商材の紹介ではなく、社会的課題の解決に向けた企業姿勢やスタンス、考え方を提示し、お客様や社会からの理解や共感を得ることを目指したコンテンツが中心となっており、イノベーションやコラボレーションをテーマにしたコンテンツを定期的に追加している。当初は、日本語/英語で展開していたが、現在では中国語、韓国語、ドイツ語など、各拠点サイトの言語でコンテンツが展開されている。

　定期的にアップデートを重ね、現在は「地球の物語の、つづきを話そう。」というサイトにリニューアルし、ブランディングにおいて大きな位置を占め続けている。

A Yokogawa brand campaign website was launched in September 2016. The site is distinct from the internally oriented Yokogawa brand site, and it has been designed to help current as well as prospective customers understand Yokogawa, and to establish a new corporate image as the company moves into its next 100 years.

An investigation of content from competing companies and other global BtoB companies was undertaken in order to study how it was presented and deployed in the digital domain, with the goal of having Yokogawa recognized as a reliable partner that collaborates with customers to meet their challenges. To achieve this, the website was not designed like a catalog, with product profiles. Instead, it presents Yokogawa's position, stance, and way of thinking with respect to meeting social challenges, and centers on content aimed at gaining understanding and empathy from customers and society. To this end, Yokogawa is adding content periodically on innovation and collaboration. The content was initially available in Japanese and English, but content in Chinese, Korean, German, and other languages has subsequently been added.

The site is updated periodically and is currently in a new iteration called "What's next for our planet? Let's make it smarter." The site continues to play a major role in Yokogawa's branding activities.

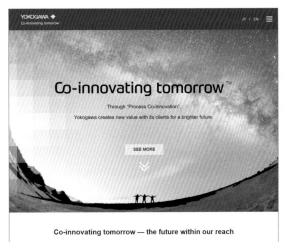

2016年9月に公開したブランド・キャンペーン・サイト

Brand campaign website
launched in September 2016

ターゲットを拡張するコミュニケーション活動の展開

Expanding Communication Activity Targets

顧客の課題解決に貢献するソリューション提供企業として、新たな市場やビジネスモデルを開拓する中で、従来の制御事業領域の企業だけではなく、ソリューションの提供を掲げる世界的なIT企業とも競合するようになった。そうした環境で成長を続けていくことは容易ではない。こうした企業は顧客の事業部の担当責任者ではなく経営層に食い込んでおり、横河電機もそこに入り込まないことには勝機はない。

コーポレートと事業部とでターゲットを分けてマーケティング活動を進めるにあたり、ブランディング推進チームが属するマーケティング本部は、具体的な製品・サービスやソリューションを紹介するのではなく、コーポレート全体の認知度を上げて、その価値をグローバルに伝える役割を担うことになった。

例えば、2017年には、「①ターゲットの拡大」と「②既存エリアの深掘り」の2つを対象として、次のような戦略を展開した。

①ターゲットの拡大

従来の制御事業領域の顧客だけでなく、広くC-レベルエグゼクティブ（経営幹部）やビジネスリーダーからのThought leadershipイメージの獲得を狙ったコミュニケーション活動を行った。

そのために活用したのが、それまではほぼ広告掲載のなかったグローバルなビジネスメディアだ。CNBC、The Wall Street Journal、Bloomberg、Reutersなどが展開するメディアへ広告を出稿するだけでなく、ブランド・キャンペーン・サイトのテーマに合わせた記事広告も制作し、各メディアに掲載された記事から横河電機のサイトへ誘導することで、企業としての考えをより理解してもらう導線もつくった。

Yokogawa provides solutions that help its customers meet whatever challenges they face. As it develops new markets and business models, the company has found itself competing not only with traditional industrial automation and control business companies, but also with global IT companies that are proclaiming their ability to provide solutions. This environment will make continued growth difficult; competitors are making inroads not with the employee in charge of business units, but at the management level. Yokogawa has no chance of winning out if it fails to rise to this competitive challenge.

In a situation where there are separate targets for corporate and business unit marketing activities, the Marketing Headquarters, which includes the Branding Promotion Team, set out to raise awareness of Yokogawa as a whole and communicate its value globally, rather than focusing on specific products, services, and solutions.

For example, in 2017, the following strategies were developed for 1) expanding branding targets and 2) exploiting existing areas more deeply.

1) Expanding Branding Targets

Communication activities aimed at positioning Yokogawa as a thought leader were directed at customers in the traditional industrial automation and control business domain, and also targeted a wide range of senior executives and business leaders in other business fields.

To this end, Yokogawa utilized the global business media, an area where the company had virtually no advertising experience. In addition to placing ads in media outlets such as CNBC, The Wall Street Journal, Bloomberg, and Reuters, Yokogawa created advertorials that matched the themes being covered on its brand campaign website. The company used these articles to direct readers to its website, thus creating a path that promoted a greater understanding of its corporate vision.

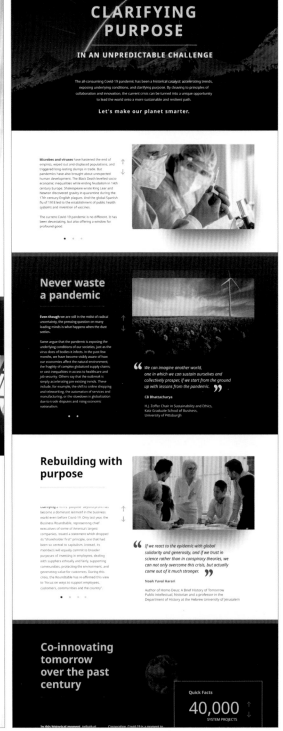

ターゲットの拡大を目的に
Bloombergと実施したタイ
アップサイト（2017年から）

Tie-up website launched
with Bloomberg to expand
branding target (from 2017)

②既存エリアの深掘り

　従来の客層に対して、持たれている良いイメージはそのまま維持するとともに、新しい横河電機の姿を知ってもらい、Technology leadershipイメージの獲得を目的とした戦略で活動を進めた。

　GMT（Global Marketing communication Team）とも協業し、制御分野の代表的な媒体を選定し、従来のような製品紹介ではなく、Co-innovating tomorrow、新しい制御事業商材のブランドやIndustrial Automationの次の世界への考え方であるIA2IA（Industrial Automation to Industrial Autonomy）などを伝える広告を出稿し、新しい横河電機の考え方やイメージの獲得を狙った。

2) Exploiting Existing Areas More Deeply

　The company pursued a strategy of promoting activities that maintained its positive image with traditional customers, while also pursuing activities that boosted awareness of the new Yokogawa and conveyed the company's technology leadership.

　In collaboration with Yokogawa's Global Marketing Communication Team (GMT), the Marketing Headquarters selected representative media in the industrial automation and control field to convey Yokogawa's new stance and image. Rather than placing conventional product advertisements, they deployed advertorials and advertisements that emphasized such themes as "Co-innovating tomorrow," new control business product brands, and our vision for a new and emerging world in which there will be a transition from industrial automation

こうした活動をしていく中で、自分たちの活動を客観的に数値化して評価することが重要だという考えの下、一定期間が過ぎたところで、CNBC、Financial Times などの協力を得て、YOKOGAWA の認知度やCo-innovating tomorrow の浸透度等に関する調査をグローバルで行った。その結果、YOKOGAWA のブランド認知、Co-innovating tomorrow の浸透度は、海外を含む各地域で着実に向上している一方で、地域間の差もあるという結果を得た。その後も毎年、調査結果を考慮しながら、ターゲットを絞ったコミュニケーション活動を展開している。

to industrial autonomy (IA2IA).

While pursuing these goals, the company believed it was important to obtain an objective, quantitative evaluation of its activities. After these activities had been deployed for a certain period, Yokogawa collaborated with CNBC, the Financial Times, and others to conduct a global survey gauging awareness of YOKOGAWA and the "Co-innovating tomorrow" concept. The results of the survey showed that, although recognition of the Yokogawa brand and awareness of "Co-innovating tomorrow" was increasing worldwide, there were also some differences among regions. Since then, Yokogawa has continued on an annual basis to deploy target-focused communication activities that take these survey results into consideration.

既存エリアの深掘りを目的に世界各国向けに同じイメージを訴求する広告を多言語で展開（2017年）

Advertisements with a consistent image were deployed internationally in multiple languages to exploit existing areas more deeply (2017)

2018年インターブランド社のアワードを受賞

Japan Branding Awards 2018

2018年11月、横河電機はそれまでのブランディングが評価され、世界最大級のブランディング専門会社であるインターブランドの日本法人インターブランドジャパンがこの年創設した「Japan Branding Awards 2018」のWinners賞を受賞した。

受賞理由として、社内を巻き込んだ体制や仕組みづくりが行われたことに加え、海外との信頼関係を構築して本社主導でありつつも自主的な取り組みをサポートして実績を築き上げた点などが挙げられた。

また、社員を重要なブランドアンバサダーとして捉え、国内外の社員への浸透活動を継続的に実施したことも、日本のBtoB企業のグローバルブランディングの取り組みとして参考になるとされた。

そして、この機会に発行されたYOKOGAWAグループの社内報『The Groupway』の2019年5月号では、「ブランドの力」と銘打った特集で、グローバルの競合企業との競争の中で価値を発揮し続けるために、なぜブランディングが必要なのかについて、改めて詳しく解説し、社員の理解の促進を図った。

In recognition of its branding to date, in 2018 Yokogawa was presented the Winners Award at the Japan Branding Awards. This event is hosted by Interbrand Japan, the Japan subsidiary of one of the world's largest branding specialists.

Reasons for the award included Yokogawa's creation of a system and framework that involved the entire company in branding, as well as its trust-building with international offices, and the way the headquarters exercised leadership while building results by supporting independent local branding efforts.

In addition, Yokogawa's view of its employees as important brand ambassadors, and its ongoing activities to promote internal brand awareness, both domestically and internationally, were acknowledged as a useful reference for Japanese BtoB companies engaged in global branding.

The May 2019 issue of The Groupway, the Yokogawa Group's internal newsletter, carried a special feature titled "The Power of our Brand." The feature, published to celebrate Yokogawa's Interbrand award, reinforced employee understanding of branding by explaining in detail why branding is important in Yokogawa's ongoing efforts to provide value in a globally competitive marketplace.

インターブランド社のJapan Branding Awardsを受賞

Yokogawa was recognized by the Interbrand Japan Branding Awards

The Groupway

MAIN CONTENTS　ブランドの力
「革新的デ　タ産業活用計画」認定

MAY 2019　**No.158**

ブランドの力

Global competitionのなかで
価値を発揮し続けるために

YOKOGAWA ◆ Co-innovating tomorrow™

Brand — 世界中で価値を発揮し続けるために求められるもの

日本の多くのお客様には、YOKOGAWAブランドが定着していますが、
世界を舞台に新たな事業領域でグローバルブランドと戦っていくためには、
ブランドの認知・理解の獲得と、競合企業には無い価値の訴求が継続して求められています。

1915 - 2000　アイデンティティの形成と日本市場におけるYOKOGAWAブランドの確立

▶ 1915
創立。創業の精神「品質第一主義」、「パイオニア精神」、「社会への貢献」は今でも受け継がれる

1920
株式会社化。「株式会社横河電機製作所」に

1927
YOKOGAWA製品の種類や数が増え、一般的に認識されるようになり、技術力、信頼性にメーカーとして責任をもつためにYEW (Yokogawa Electric Worksの略)を商標登録。当時の日本のメーカーは丸形の商標が多かったがドイツメーカーに多く見られる角形を採用した

YEW 横河北辰電機

1983
北辰電機製作所と合併し、「横河北辰電機株式会社」に社名変更。当時の経営者が「合併比率は資産や株価から決めたが人財と技術は対等だ」との考え方を簡潔などに徹底し、社名も両社を冠するものに

1986
CI (Corporate Identity) 導入。社名を横河電機に変更し、海外の人でも覚えやすいよう、現在のシンボルマークとブランドロゴに変更

YOKOGAWA

1988
合併の仕上げとして企業理念を制定

YOKOGAWAは計測と制御と情報をテーマにより豊かな人間社会の実現に貢献する
YOKOGAWA人は良き市民であり勇気をもった開拓者であれ

YOKOGAWAの考えるブランドとは

ブランドとは、一般的にはその企業の企業価値（製品、サービス、ソリューションなど）を他社が提供するそれらのものと区別するための「企業としての約束」と考えられています。提供する側にそのブランドに対する明確なビジョンがあって、社員によるその共有と確信と実践があり、その価値はお客様、市場、社会に認知、理解、評価していただくことで初めて成立します。

そして、ブランドを醸成、強化するブランディング活動には、一般的な広告キャンペーンとは異なり、ビジョンをもった中長期的な活動が求められます。

ブランディングの必要性について特集で詳しく解説したYOKOGAWAグループの社内報『The Groupway』（2019年5月号）

The May 2019 issue of the Yokogawa Group's internal newsletter, The Groupway, included a special feature explaining the importance of branding in detail

顧客と社員を対象にして定期的に調査を実施
Periodic Customer and Employee Surveys

ブランディングというのは、ややもすると担当部署だけが積極的に進める取り組みになってしまい、社員や顧客がついてこられないという結果になりがちである。そうしたことを避けるために、ブランディング推進チームは、常に結果を確認しながら戦略を展開することを怠らなかった。

「こうした活動は実行したことによる効果を示すことが重要です。私たちは効果を定量的に評価することを重要視しており、同時に定性評価も行い、お客様や社員におけるブランド浸透度を測定しています」とメンバーは語る。

具体的には、顧客や社員を対象とした調査を実施している。

顧客に対しては、2013年、2017年、2020年にグローバルで比較的大規模な自主調査を実施し、2017年、2020年の調査ではコーポレート・ブランド・スローガンであるCo-innovating tomorrowに対する認知も調査対象とした。既存顧客に対しては、Co-innovating tomorrowに対する認知・理解の獲得を目標に設定していた。2017年までに20%、2018年までに30%、2020年までに50%を目標としていたが、2018年の調査の段階で約40%、直近の2020年の調査では約60%が、Co-innovating tomorrowを認知しているという結果を得ている。

社内に対しても2016年以降、コーポレート・ブランド・スローガンに対する意識調査を実施している。当調査では、コーポレート・ブランド・スローガンの日常業務における体現度を聞いているが、2018年度の調査では、「常に意識している」が36%、「業務によって意識している」は42%という結果になった。その後の調査では年々意識が高まっていることがうかがえ、2022年度の調査では、それぞれ52%、36%となっている。

Branding tends to be pursued aggressively only by the department that has responsibility for them. As such, employees and customers are often left behind. To avoid such outcomes, the Branding Promotion Team makes sure that it always monitors results when it deploys a strategy.

"It is important to verify the effectiveness of our activities. We regard quantitative evaluation of effectiveness as very important. At the same time, we conduct qualitative evaluations to measure brand awareness among customers and employees," said a team member.

The team performs these measurements by surveying customers and employees.

The company conducted relatively large global customer surveys in 2013, 2017, and 2020. The 2017 and 2020 surveys also gauged awareness of the "Co-innovating tomorrow" corporate brand slogan. For existing customers, targets were established for gaining awareness and understanding of "Co-innovating tomorrow." These targets were 20% by 2017, 30% by 2018, and 50% by 2020. The 2018 survey showed that approximately 40% of all respondents were aware of the corporate brand slogan, and in the 2020 survey this figure increased to approximately 60%.

Since 2016, the company has also been surveying employees to gauge awareness of the corporate brand slogan and to judge the degree to which the corporate brand slogan is embodied in their day-to-day operations. In the fiscal year 2018 survey, 36% of respondents reported always being aware of the corporate brand slogan, while 42% reported that this depends on what work they are engaged in. Subsequent surveys have shown awareness rising year by year, with 52% and 36% awareness levels, respectively, in the survey of fiscal year 2022.

コーポレート・ブランド・スローガンを「常に意識している」社員が増加したことは、ブランディングを"自分ごと"として考える姿勢が社内に定着したことを示していると考えてよいだろう。この調査結果を受けて、ブランディング推進チームはそれまでの活動に強い確信と自信を得ることができた。

2017年から2018年にかけては、海外へのブランディングの浸透と定着において重要な時期であったと同時に、継続的な活動が一つの成果をもたらした時期であったとも言えよう。

コーポレート・ブランド・スローガンへの認知の高まりを受けて、横河電機のブランディングは、ステークホルダーとの共創という新たな局面へと進んでいった。

The increase in the number of employees who are always aware of the corporate brand slogan can be considered an indication that the mindset that branding is of personal relevance is taking root among employees. These results have reinforced the Branding Promotion Team's confidence and trust in their activities to date.

The period from 2017 to 2018 was critical in the building of awareness and establishment of branding internationally. During this period, the Branding Promotion Team delivered results through their ongoing activities.

With the growing awareness of the corporate brand slogan, Yokogawa's branding has progressed to a new phase of co-creation with stakeholders.

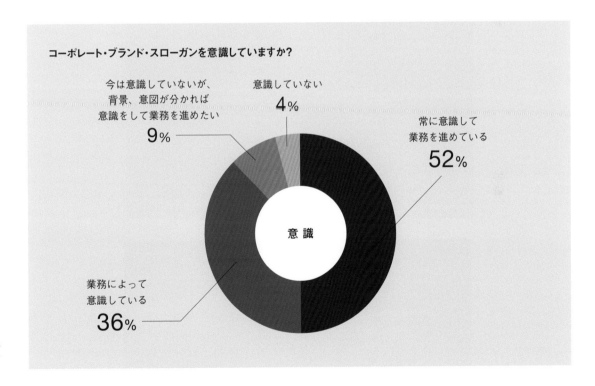

コーポレート・ブランド・スローガンを意識していますか?

今は意識していないが、背景、意図が分かれば意識をして業務を進めたい
9%

意識していない
4%

常に意識して業務を進めている
52%

意識

業務によって意識している
36%

2022年度(2023年2月〜3月)に、YOKOGAWAグループ全社員を対象に実施した社内浸透度調査の回答結果

Results of a fiscal year 2022 internal penetration survey targeting all Yokogawa Group employees (conducted February to March 2023)

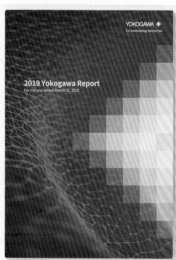

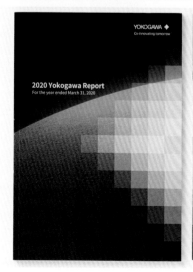
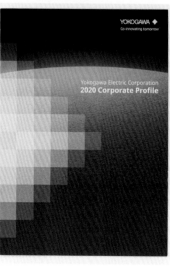

Brilliant Gridを表紙に使った
統合報告書と会社案内

Cover of integrated report
and corporate profile featuring the Brilliant Grid

Contribution

事業構想大学院大学 学長

田中 里沙 氏

BtoB企業が持つ技術や知財を共創の軸に
ブランディングで暗黙知を形式知化する

　BtoB企業はBtoC企業に比べれば生活者との接点が多くなく、新規事業展開や大型の設備投資、M&Aなどのニュースがない限りはなかなか注目されにくい。加えて、社会からは誠実さや信頼を求められるため、派手な行動を避ける意識が強く、それがブランド認知度向上における課題となっている。

　しかし、BtoB企業も社会とともにあるのだから、世の中に積極的にアピールをして多様な意見を取り入れるようにすべきだと考えている。外部の人や組織との共創はイノベーションの創出にもつながる。では、具体的にどうすればよいのか。

　私はジャパン・ブランディング・アワードの設立時から審査副委員長を務めており、2018年の横河電機の受賞は今も強く印象に残っている。それは、ブランド価値を高める戦略において社内の巻き込み方をはじめ、体制構築や仕組みづくりに注目すべき点があったからだ。

　ブランディングにおいては、よく「社員とのエンゲージメント構築が難しい」と言われる。だが同社は、関係部署が密接に情報交換を行い、エンゲージメントのシナリオづくりに力を入れ、かなりの時間を割いて周到な下準備をしていたことがうかがえた。

　企業の中には宣伝、広報、経営企画部門などがバラバラに動くために、新聞報道や広告キャンペーンを見て初めて社員がその内容を知る場合があると聞く。これは、社員のやる気を著しく下げる原因となることもある。そのため同社では、あらかじめ公開情報を社内サイトで周知しているという。社員からは「こんな広告を見た」「あそこに人が集まっていた」という報告が続々と寄せられるそうだ。こうした事象はマーケティング担当者冥利に尽きると言えよう。何より全社員がブランディング担当者と同じ心持ちでいることが同社の強さなのだと感じる。

　国内のグローバル企業の多くが苦労しているのが、海外の社員へのエンゲージメントだ。日本とは考え方や価値基準、コーポレートブランドとの距離感も違う。そこで、各地域の社員や顧客を引き付けるために重要になるのが、理念やパーパスなど、よりどころとなるメッセージに一本筋が通っていることだ。そう考えると同社のスローガンやメッセージは、異なる国や文化において理解しやすい言葉で端的に示されているところが効果的であると思う。意識の浸透は簡単ではない。「スポンジに水が染み込むように浸透していく」ためにはそれなりの工夫や議論が不可欠だ。

　同社の近年の活動の中では「社員がコミットする中計」という言葉に感服した。中期経営計画に社員が主体となって関与するのは理想ではあるが、そう簡単にできるものではない。さらに「社員みんなでつくるパーパス」には現場力の強みを感じた。パーパスは2021年ごろから世界的に流行し、中にはコンサルタントに依頼してつくるケースも見受けられる。それに対して同社では、パーパス策定の過程において「自分自身は何のために仕事をしているのか」「横河電機を通して社会とどうつながっているか」を社員一人ひとりに問うたことが画期的と言えよう。近年は「参加」がキーワードになっているが、参加したくなる環境を生み出して、自主性を触発する言葉や仕掛けが数多くあるように感じられた。これらをヒントにして、多くの企業が自分の会社にしかできない参加の方法を考えてほしい。

　日本のBtoB企業は素晴らしい技術や知的財産を持っているのに、残念ながらそれらが世の中に知られていないことが多い。社内では当然と受け止められていても、外から見れば決して当たり前ではない。自分たちが持つ技術や知財をもっとアピールすれば、私たちのような大学や自治体といった組織との共創が生まれ、ダイナミックなイノベーションを起こすきっかけにもなるだろう。企業が有する暗黙知を形式知化することはイノベーションを生み出すために非常に重要であり、それに寄与するのがブランディングなのである。

Profile

マーケティング、コミュニケーションを専門とし、雑誌「宣伝会議」編集長を経て、株式会社宣伝会議取締役メディア・情報統括。内閣府政府広報アドバイザー、中央環境審議会、地方制度調査会等の委員。環境省クールビズネーミング、東京2020エンブレム、G7、G20サミットロゴマーク選考委員等を務めた。2012年より学校法人先端教育機構 事業構想大学院大学教授、2016年学長に就任。2021年より国立大学法人三重大学理事（広報・ブランディング担当）。

Contribution

President, the Graduate School of Project Design

Risa Tanaka

Converting Tacit Knowledge into Explicit Knowledge by Branding BtoB Company Technology and Intellectual Property as a Co-Creation Axis

Compared to BtoC, BtoB companies have less contact with consumers, and without news of new business development, major capital investments, or M&A activity, it is difficult for the latter to attract attention. Furthermore, society values sincerity and trust highly, and as such BtoB companies are very reluctant to engage in behaviors designed to draw attention, which poses a challenge from a brand awareness enhancement standpoint.

However, since BtoB companies operate within society, I believe they should appeal actively and strive to incorporate diverse opinions. Co-creation in collaboration with outside individuals and organizations will also promote innovation formation. How, then, can this be accomplished in practice?

I have served as vice chair of the Japan Branding Awards judging committee since its establishment, and I still remember Yokogawa Electric Corporation's capture of the 2018 award vividly. The way the company mobilized internal involvement in its brand value enhancement strategy, as well as its system building and framework creation, were noteworthy.

It is often said that building engagement with employees with respect to branding is challenging. However, Yokogawa evidently spent considerable time preparing carefully for the project, with relevant departments exchanging information closely and focusing on creating engagement scenarios.

I am told that the advertising, public relations, and management planning departments of some companies work independently, and employees may not be aware of the content of an article or an ad campaign until they actually encounter it. Such experiences can significantly impact employee motivation. For this reason, Yokogawa discloses this type of information in advance on its internal website. Employees submit information on ads they have seen or that attract the public. Such information represents a great opportunity for the company's marketing professionals. Above all, the fact that all of its employees share the same mindset as the branding team is a major strength of the company.

Many global Japanese companies struggle to engage their international employees. The mindset, values, and distance from the corporate brand of these employees differ from those of employees in Japan. As such, to attract employees and customers in diverse regions, it is important that the company's messaging, which serves as a reference for its philosophy and purpose, be consistent. In this sense, Yokogawa's slogans and messages are effective because they are presented in simple terms that are understood easily in diverse countries and cultures. Spreading awareness is no easy thing. A certain degree of planning and discussion is required to penetrate awareness, like water soaking into a sponge.

Of the company's recent activities, I was impressed by the phrase "mid-term business plan that employees commit to." While it is ideal for employees to be proactively involved in mid-term business planning, doing so is not easy in practice. Furthermore, I sensed frontline strength in the phrase "purpose created by all employees." The "purpose" concept has become popular worldwide since around 2021, and in some cases consultants have been commissioned to create them. Yokogawa's approach was to ask each employee, in the process of developing their own purpose, what their personal reason for working is, and how they are connected to society through the company, and I think this was truly innovative. In recent years, participation has become a key word, and I feel that there are many words and devices that create environments that promote the desire to participate and inspire one's autonomy. I hope more companies will take these hints and devise their own unique ways of participation.

While Japanese BtoB companies have wonderful technologies and intellectual property, these are, unfortunately, often not well known to society. They may be taken for granted within the company, but they would never be taken for granted outside it. If such companies could better promote their technologies and intellectual property, they could likely co-create with organizations—universities like ours, local governments, and so forth—to create opportunities for dynamic innovation. The conversion of tacit knowledge into explicit knowledge by the enterprise is extremely important for generating innovation, and branding contributes to this process.

Profile

As a marketing and communications specialist, she served as managing editor of Sendenkaigi magazine and Director, Media and Information, of Sendenkaigi Co., Ltd. She is currently a governmental publicity advisor to the Cabinet Office, and is a member of the Central Environment Council and the Local Government System Research Council. She also served on the Cool Biz Naming Committee for the Ministry of the Environment, the Tokyo 2020 Emblem Committee, and the Logo Screening Committees of the G7 and G20 Summits. In 2012, she became a professor at the Graduate School of Project Design, Advanced Academic Agency, and in 2016 was appointed president of the school. In 2021, she became a director (responsible for PR and branding) of Mie University.

Chapter 4

展 開
Deployment

製品、サービス、ソリューションの
領域におけるブランディング

Branding in the Areas of Products,
Services, and Solutions

コーポレートブランドから製品ブランドへの展開
Deploying the Corporate Brand to Products

コーポレートブランディングが順調に進んでいく一方で、大きな課題が残されていた。それは、製品ブランドおよび製品名称の体系整理である。横河電機の制御事業は、長い歴史の中で1000を超える製品やサービスの名称が生まれており、それらは体系化されているとは言えなかった。

数多くの製品やサービスの名称が使われていたのは、計測機器や制御システムの提供が事業の中心だった時代の名残で、事業拡大により製品が増えた上、各開発組織が独自に名称の判断を行っていたこと、また、製品のコモディティ化に伴う整理や再体系化がされてこなかったことの結果だった。包括的なソリューションを通してお客様へさらなる価値の創出・提供を目指している中、グローバル化や課題解決型ビジネスへの移行、技術進歩による製品機能拡張やデジタル化など、①ビジネス環境の変化、②顧客・市場の変化、③技術の変化という3つの大きな変化によって、制御ビジネスおよび企業価値向上への貢献が必須となってきた。

こうした変化を視野に入れ、ブランディング推進チームは、製品、サービス、ソリューションの領域においても今後のビジネス環境を見据えた効果的・効率的な制御事業としてのネーミング方針を策定する必要があると判断。2017年には組織を横断したネーミングプロジェクトが発足した。

Yokogawa's Electric Corporation corporate branding was progressing smoothly, but a major challenge remained: the systematic organization of the company's product brands and names. Over the long history of Yokogawa's industrial automation and control business, more than 1,000 product and service names had been created, and those names could not be regarded as well-organized.

The plethora of product and service names was a legacy of the era when the sale of measurement devices and control systems was the company's core business. As Yokogawa grew, so did its product lineup, and individual development teams decided product and service names independently. In addition, as some products became commodified, adjustment and re-systematization of the product line was neglected. As Yokogawa aims to create and deliver even more value to its customers through comprehensive solutions, three major changes have made it imperative to contribute to the control business and to increase corporate value. These changes are 1) changes in the business environment, 2) changes in customers and markets, and 3) changes in technology, and these include globalization, the shift to problem-solving businesses, and the expansion of product functions and digitalization in the wake of technological advances.

With these changes in mind, the Branding Promotion Team decided it was necessary to formulate an effective and efficient control business naming policy that was designed to meet future business demands, in the areas of services and solutions as well as products. In 2017, the team launched a cross-organizational naming project.

長い歴史の中で増えていた製
品やサービスブランドのロゴ一例

Some examples of product/
service brand logos that
proliferated over Yokoga-
wa's long history

Deployment

製品、サービス、ソリューションの包括ブランド「OpreX」の誕生

The Birth of OpreX: A Comprehensive Products, Services, and Solutions Brand

製品やサービスのブランドが体系化されていないことは、以前から課題として認識されていた。しかし、各事業部の方向性が異なり、十分な情報共有や連携がされずにそれぞれが活動していたため、組織を横断する形でのブランド体系化は困難だったのだ。

「この機会を逃せば、競合相手に置いていかれてしまいます。ブランディングの最後のピースを埋めるという認識の下、包括ブランドの策定に取り組む決意をしました」とメンバーの一人は語る。

まず行ったのは、グローバルの競合相手との比較である。その結果、横河電機の事業規模をはるかに超える有名なグローバル企業であっても、製品ブランドやシステムブランドは、ブランド価値を集約するべく、数を絞り込んでいることが判明した。一方、当時の横河電機には前述のように1000を超える名称があった。

これでは顧客や市場に混乱を与えかねない。ソリューションビジネスのさらなる展開に適した商材の体系化やポートフォリオ化も必要になってくる。そこで、制御事業に関する製品、サービス、ソリューションを包括するブランドを新たに立ち上げることになった。新しいブランドの条件は、「同一の製品は世界で同じ名称に統一されていること」「世界中の人が容易に理解できる商材名であること」「グローバルビジネスにおける商標侵害リスクの低減や、登録・維持費用を削減できること」だった。

ブランディング推進チームが中心となってブランド名称を検討し、2018年5月に、制御事業ポートフォリオ全般を包括する「OpreX」(オプレックス)ブランドを策定。OpreXとは、"Operational""Excellence""DX technology"という3つの要素を統合した造語だ。

中でも "X" を大文字で表記し、強いメッセージ性を持たせた。このXには次の3つの意味が込められている。

The fact that product and service brands were not systematized was recognized as an issue. However, each business unit was pursuing a different direction, without adequate information sharing and coordination. As such, cross-organizational brand systematization had become difficult.

"If we failed to seize this opportunity, our competitors would leave us behind. We were determined to work toward establishing a comprehensive brand, and were aware that we were completing the final element of our branding," said a Branding Promotion Team member.

The team's first step was a comparative study of global competitors. They found that even renowned global companies, ones far exceeding Yokogawa's business scale, had focused on their product and system brands to concentrate their brand value. Yet, as noted, at the time of the comparison study, Yokogawa maintained more than 1,000 product and service names.

This could confuse customers and the market. Systematizing the names of solutions business products and organizing them into portfolios in a manner appropriate for further business expansion became a necessity. As such, the team resolved to launch a comprehensive new brand encompassing all industrial automation and control business products, services, and solutions. The new brand would have to satisfy the following conditions: unified worldwide naming for each product, product names that are easy to understand for people around the world, and a lower global risk of trademark infringement, with reduced registration and maintenance costs.

In May 2018, the Branding Promotion Team took the lead in considering the brand name, and OpreX was adopted to encompass the entire industrial automation and control business portfolio. OpreX is a blend word or portmanteau that is derived from the expressions operational, excellence, and DX.

The capital letter X conveys a strong message, and has the following three connotations.

①「交差」の象徴としてのX

システム化、ネットワーク化、領域の横断、ステークホルダーとの共創を意味する

②「ex-」「trans-」という接頭辞を意味するX

Excellent（卓越した）、Transformation（変化）を連想させる

③未知数を示すX

明るい明日を実現するために顧客が体験したことのない有益な技術や価値を提供する

さらにこの①～③は、コーポレート・ブランド・スローガン「Co-innovating tomorrow」の「①Co：ステークホルダーとの共創」「②innovating：変化」「③ tomorrow：明るい明日」にもつながっている。

1) X symbolizing Intersection
Systemization, networking, cross-domain, co-creation with stakeholders

2) The prefixes ex and trans
Association with the words excellence and transformation

3) An unknown variable that signifies future potential
Delivery of unprecedented, beneficial technologies and customer value to realize a brighter tomorrow

Moreover, the three connotations link to the "Co-innovating tomorrow" corporate brand slogan: 1) Co: Co-creation with stakeholders, 2) innovating: transformation, and 3) tomorrow: a brighter tomorrow.

制御事業の製品・サービス・ソリューションの包括ブランド OpreX発表（2018年）

Announcing OpreX, a comprehensive brand offering industrial automation and control business products, services, and solutions (2018)

コーポレートブランディングにリンクした製品ブランディング

Product Branding Linked to Corporate Branding

OpreXは5つのカテゴリーで構成されている。その下位に25のファミリーが設けられ（当時）、約1000種類の名称はその中に分類された。つまり、「OpreX－カテゴリー名－ファミリー名－製品名－モデル名」によりすべての製品を分類するネーミングルールを制定したのである。

商標登録は全製品の先頭に冠するOpreXのみとしたため、知的所有権の登録・維持に関する費用の大幅な削減にもつながった。

またロゴは、Co-innovating tomorrowと同じくPrometo書体をベースとしている。これには、多くの製品名をこの書体で統一し、企業としての一貫したブランドを顧客や市場に示すという意図がある。言うまでもなく、ロゴの使い方に関してもコミュニケーション・デザイン・ガイドラインを制定し、グローバルレベルでの順守を求めている。

OpreXの発表後は認知度を向上させるために、Webサイトや雑誌などのメディアを使い、グローバルでの展開や浸透に努めた。新しい製品ブランドの紹介ではなく、コーポレートブランディングの一環という意味を込めて、あえて製品の写真は入れずに、OpreXのロゴを前面に出す広告を展開した。

The OpreX brand encompasses five categories and (initially) a total of 25 family subcategories with approximately 1,000 product names, using the naming rule "OpreX - category name - family name - product name - model name."

Since all product names begin with the common OpreX brand, a significant reduction in trademark-registration and other IP-related costs has been realized.

In addition, the logo is based on the same Prometo typeface used for "Co-innovating tomorrow." The goal is to unify many of the company's product names with this typeface and present a consistent corporate brand to customers and the market. Naturally, the company has created Communication Design Guidelines for logo use, and requires adherence to these guidelines on a global level.

Following the announcement of the OpreX brand, the company worked to enhance awareness by deploying and disseminating it on websites, magazines, and other media around the world. Rather than introducing it as a new product brand, advertisements prominently featured the OpreX logo as a corporate brand, unaccompanied by product photos.

OpreXのロゴを前面に打ち出
した広告を展開（2018年）

Ad placements prominently
featuring the OpreX logo
(2018)

セットで携行できるブローシャー
とフォルダー

Brochure and folder to carry
it in

ネーミングプロジェクトのコミッティー化に踏み出す
Formation of a Naming Project Committee

　OpreXの策定は各事業部の理解を得て順調に進んだものの、ブランディング推進チームがそこにとどまることはなかった。過去にも何度か統合ブランド策定の試みはあったが、それが成功しなかった理由を分析していたためである。

「従来は、ブランド制定のプロジェクトが一段落したらプロジェクトは解散していました。しかし、今回は一過性で終えてはならないと結論付けました」とメンバーの一人は語る。

　そこで、製品、サービスやソリューションのネーミングルールにもコーポレートでのマネジメントが必須と判断して、ビジネス環境の変化に永続的に対応するためのコミッティー化に踏み出した。

　こうしてネーミングプロジェクトは、2019年に「OpreXネーミングコミッティー」に昇格した。各事業部の意見をとりまとめて集約できるポジションの管理職をメンバーに据え、ネーミングにガバナンスを効かせる体制にしたのである。新製品を発売する際には、コミッティーにネーミングの申請をして認可を得る仕組みとした。さらに2021年には、制御事業関連だけでなく、他の事業分野や全事業に関係するプラットフォームの名称等も協議するために、「YOKOGAWAネーミングコミッティー」へと発展。今後もビジネス環境の変化とともに進化していくだろう。

Although the OpreX launch went smoothly thanks to the cooperation of business units, the Branding Promotion Team was not content to stop there. This was because they had analyzed previous unsuccessful attempts to establish an integrated brand.

"Previous projects to establish a brand were disbanded once a goal was reached. But we concluded that this time, the project would not be a one-off effort," said a team member.

The team determined that it was essential for products, services, and solutions naming rules to be managed at the corporate level. They therefore decided to establish a committee that would provide ongoing oversight amidst a changing business environment.

Thus, in 2019, the naming project was upgraded with the establishment of an OpreX Naming Committee. The committee was composed of managers who could gather and summarize the views of people in different business units, and a governance framework was introduced for the naming process, with naming applications and committee approval required when launching a new product. In 2021, this committee was renamed the Yokogawa Naming Committee, and assumed responsibility not only for naming related to the industrial automation and control business, but also for other business domains, and the naming of shared business platforms. In response to a changing business environment, the committee will likely continue to evolve.

Chapter 5

発展
Expansion

社員参画による「パーパス」策定と
ブランディングの新たな展開

Formulation of the Purpose Statement
through Employee Participation and New Launch of Branding

「地球の物語の、つづきを話そう。」をタグラインとした広告展開

"What's Next for Our Planet? Let's Make It Smarter." as the Tagline of the Advertising Campaign

2019年に入ると、ブランディングは新しい局面に入っていく。コーポレート・ブランド・スローガン「Co-innovating tomorrow」を基本思想として、地球の未来についてステークホルダーとの共創を目指すフェーズに向かっていったのである。

その取り組みに向けて必要とされたのは、次の3つのメッセージの発信であった。

・横河電機は制御事業のみの会社ではなく、ライフサイエンス、バイオエコノミーなどの新しい事業領域でも、顧客の成長と変革を支援する企業である。

・2050年に向けて目指す社会の姿を定めたサステナビリティ目標「Three goals」を掲げ、社会課題の解決、SDGsに応える企業である。

・エネルギーやマテリアルでのゼロエミッション、循環型経済といった社会課題を、事業を通じて解決していく企業である。

折しも、2019年のリクルートシーズンを迎え、横河電機では新しい方向性を模索していた。そこで、ブランディング推進チームは、「地球の物語の、つづきを話そう。」をタグラインとして、直接のターゲットであるリクルート活動をしている学生に加えて、広くその親世代、潜在的な顧客向けにコーポレートブランドの向上を目指すメッセージを発信する広告を出稿する。

2019年3月に展開された広告は、従来型のリクルー

In 2019, the company's branding efforts entered a new stage. With the "Co-innovating tomorrow" corporate brand slogan as the fundamental concept, the company proceeded to a phase in which the focus was on co-creating the future of the planet with stakeholders.

For that initiative, the company needed to transmit the following three messages:

•Yokogawa is more than an industrial automation and control company; it supports growth and innovation by customers that are engaged in new business fields such as life sciences and the bio-economy.

•Yokogawa is a company that is working to address societal challenges and achieve the Sustainable Development Goals, and is guided in this by its "Three goals" for sustainability by the year 2050.

• Yokogawa is a company that, through its businesses, resolves challenges such as the need for zero emissions and a circular economy in the energy and materials business segments.

As the 2019 new employee recruiting season approached, Yokogawa was looking for a new direction. And so, with "What's next for our planet? Let's make it smarter." as the tagline, the Branding Promotion Team placed ads to strengthen the corporate brand image with university job applicants, targeting not only students but also members of their parents' generation and potential customers.

「地球の物語の、つづきを話そう。」TVCF

"What's next for our planet? Let's make it smarter." TVCF

話そう、
バイオが育む
未来に
ついて。

穏やかな人生、豊かな社会を実現するために挑むべき課題は無数にある――。
医療や創薬、食品、バイオ燃料など、バイオテクノロジーの
発展に挑戦するYOKOGAWAのライフイノベーション事業は、
多くの社会的事業に向けて、これまで培ってきた
計測、制御、情報の技術を、応用して提供。
基礎研究から製造、物流・サービスとやバリューチェーン全体での
生産性革命を実現し、お客様との共創を通じて、
人や社会への貢献という、創業以来の理念を実現していきます。

地球の物語の、つづきを話そう。

YOKOGAWA ◆
Co-innovating tomorrow™

横河電機株式会社 横河計測株式会社 横河商事株式会社 横河ソリューションサービス株式会社 横河マニュファクチャリング株式会社 横河レンタ・リース株式会社 エム・アイ・エス株式会社
Co-innovating tomorrow および Co-innovating tomorrow ロゴは横河電機の登録商標です。

話そう、
ひとつの細胞が
持つ宇宙に
ついて。

「やわらかいもの」を測る――。
そのテーマを出発点にYOKOGAWAが開発した高解像度共焦点スキャナユニット
が、細胞の活動や薬などに対する反応のメカニズムを明らかにし、
病理の解明や新薬の薬効、副作用が発生するしくみを解析する
ソリューションへと応用されています。
再生医療や、遺伝子レベルでの分析で患者様4人に最適な治療を施す
医療への期待が高まるなか、私たちは、お客様との共創を通じて、
積極的に最新医学の研究開発支援に取り組んでいます。

地球の物語の、つづきを話そう。

YOKOGAWA ◆
Co-innovating tomorrow™

横河電機株式会社 横河計測株式会社 横河商事株式会社 横河ソリューションサービス株式会社 横河マニュファクチャリング株式会社 横河レンタ・リース株式会社 エム・アイ・エス株式会社
Co-innovating tomorrow および Co-innovating tomorrow ロゴは横河電機の登録商標です。

話そう、
新しい産業の
プラットフォームに
ついて。

Automation（自動）からAutonomy（自律）へ――。
AIやIIoT、ロボティクスを活用した
人の手介在のない自律制御で運転する工場。
そこでは完全な自動化、あるいは無人化を実現することができます。
YOKOGAWAは、さまざまな生産・製造の現場で培ってきた経験をもとに、
計測、制御、情報の技術と最新のITを融合させ、Autonomyへの移行を推進し、
産業界全体のトランスフォーメーションをサポート。
お客様と"ものづくり"の新たなフロンティアに挑戦します。

地球の物語の、つづきを話そう。

YOKOGAWA ◆
Co-innovating tomorrow™

横河電機株式会社 横河計測株式会社 横河商事株式会社 横河ソリューションサービス株式会社 横河マニュファクチャリング株式会社 横河レンタ・リース株式会社 エム・アイ・エス株式会社
Co-innovating tomorrow および Co-innovating tomorrow ロゴは横河電機の登録商標です。

話そう、
私たちの
「働き方改革」に
ついて。

社員一人ひとりの個性が集まってYOKOGAWAになる――。
だからこそ社員に期待することがあります。それは働の力を強くすること、
ボーダーを超える挑戦を恐れないこと、そして社会に対して貢献すること。
そうすれば、やりがいや働きがいは、自ずと見えてくる！
「働き方」の質が問われる今、自らの意志をもって成長し、
行動に移す社員の挑戦を後押しする。
それが、これからの"YOKOGAWA人"を育てていくにつながり、
お客様に対する最大のソリューションになるはずです。

地球の物語の、つづきを話そう。

YOKOGAWA ◆
Co-innovating tomorrow™

横河電機株式会社 横河計測株式会社 横河商事株式会社 横河ソリューションサービス株式会社 横河マニュファクチャリング株式会社 横河レンタ・リース株式会社 エム・アイ・エス株式会社
Co-innovating tomorrow および Co-innovating tomorrow ロゴは横河電機の登録商標です。

「地球の物語の、つづきを話そ
う。」広告クリエイティブ

"What's next for our planet?
Let's make it smarter." ad
creative

横河電機なのに iPS細胞について 考えている。

**今も社名に「電機」が残るのは、
創業時の開拓者精神を継承しているから。**

もっと精密に、もっと革新的に。
その思いから、日本で初めて電気計器の実用化に成功した横河電機。
以来、「電機」にとどまらない事業領域に挑み、革新を起こしてきました。
計測、制御、情報そしてライフイノベーションといった領域において、
ビジネスの変革のみならず、社会的課題の解決にも深く関わってきました。
例えば、今日、私たちのプラントにおける高度な計測技術は、
3D観察が可能な共焦点スキャナに応用され、
細胞内小器官の動態観察を可能にするなど、ライフサイエンスの発展にも寄与しています。
さらに、省エネルギー性に優れたバイオプロセスの実現に取り組み、
再生可能な生物由来の資源活用を推進していきます。
これからもパイオニア精神をもって新しい事業領域に挑み、
美しく豊かな地球を未来世代へつないでいきます。

地球の物語の、つづきを話そう。

YOKOGAWA ◆

Co-innovating tomorrow™

Co-innovating tomorrow および Co-innovating tomorrow のロゴは横河電機の登録商標です。

横河電機株式会社

キャンペーンタグライン「地球の
物語の、つづきを話そう。」を初
めて使用した新聞広告

First newspaper advertisement with the campaign
tagline, "What's next for
our planet? Let's make it
smarter."

トメディアに加えて、様々なSNSメディアにも出稿して若年層にアピール。加えて日本経済新聞等にも出稿した。そこでは、「横河電機なのにiPS細胞について考えている。」「YOKOGAWAは今、地球の相続について考えている。」という刺激的なコピーを使って、ライフサイエンスやバイオ関連ビジネスなど、新たな領域に事業を広げていることを示した。

また、2019年の大みそか、そして続く元日には広告展開の第2弾として、2日連続で日本経済新聞に出稿。交通広告、ソーシャルメディアにも展開した。その後、リクルート活動を意識し事業内容をより理解してもらうため「話そう」をキーフレーズとした広告も展開。特にソーシャルメディアでは前回の結果を受け、動画の掲載へと活動を広げた。いずれも学生だけでなく、親世代も意識したメッセージを掲載し、「地球の物語の、つづきを話そう。」の内容を具体的に理解できるようにした。

一方、2016年9月に公開したCo-innovating tomorrowのブランド・キャンペーン・サイトは、「地球の物語の、つづきを話そう。」サイトとして、地球環境やサステナビリティを強く意識したコンテンツを収録し、発信していくようになる。

こうした広告展開も含めた取り組みに加えて、人財部門の採用活動の結果、2020年度の新卒採用の応募数は前年比30%増となり、その半数が女性の応募となった。また、バイオエコノミー、ライフサイエンスをバックグラウンドに持つ応募者も大幅に増加した。日本経済新聞の調査によると、「横河電機のイメージが変わった」というコメントが多く寄せられたという。社内からも予想以上の反響があり、後のYokogawa's Purpose策定のプロローグとなっていく。

In addition to traditional recruiting media, the advertising campaign that was launched in March 2019 also reached out to young people via posts on various social media platforms. Ads were also placed in newspapers such as the Nikkei. Using intriguing copy such as "Yokogawa Eyes iPS Cells" and "Yokogawa Protects the Planet's Heritage," the content showed that the company is expanding its business into new fields such as the life sciences and bio-related businesses.

As the second installment in the advertising campaign, the team placed ads in the Nikkei two days in a row, on New Year's Eve 2019 and New Year's Day 2020. The ads also appeared in public transportation and on social media. After this, the team ran an advertising campaign with "What's next?" as the key phrase, with the goal of raising awareness about the company's recruiting activities and understanding about its business. In particular, social media activities were expanded, based on previous results with those platforms, with the addition of more videos. Messages were addressed not only to students but also to their parents' generation, giving more detailed content for "What's next for our planet? Let's make it smarter."

Meanwhile, the "Co-innovating tomorrow" brand campaign site that had been set up in September 2016 was updated to cover "What's next for our planet? Let's make it smarter." The aim was to gather together and share content that promoted a strong awareness of environmental concerns and the need for sustainability

As a result of the HR department's hiring activities along with initiatives such as these advertising campaigns, the number of new graduates applying to work at Yokogawa in fiscal year 2020 increased by 30 percent over the previous year, and half of these were women. Furthermore, the number of applicants with backgrounds in bioeconomy and the life sciences increased significantly. According to a Nikkei survey, many commenters mentioned that their image of Yokogawa had changed. The impact within the company also exceeded the team's expectations, serving as a prologue for the formulation of Yokogawa's Purpose.

「地球の物語の、つづきを話そう。」ブランド・キャンペーン・サイト

"What's next for our planet? Let's make it smarter." brand campaign website

次期中期経営計画策定とパーパス策定の背景
Background to the Formulation of the Next Mid-term Business Plan and Formulation of the Purpose Statement

2020年は、2021～2023年を対象年とする次期中期経営計画を検討する年であった。中計は、社内におけるブランディングにとって非常に重要な機会であるため、ブランディング推進チームは過去10年間で3回の中期経営計画策定に積極的に関わってきた。

中期経営計画『Transformation 2017』では、社員のベクトルを束ねることを目的に、同時に発表された長期経営構想にあわせ、コーポレート・ブランド・スローガンCo-innovating tomorrowを制定。次の中期経営計画『Transformation 2020』では、一歩進んで社員全員の中期経営計画と位置づけ、その理解を促進し、ベクトルを合わせるために、社長・役員から社員に向けて対面で内容を説明した。

そして、2021年に策定された中期経営計画『Accelerate Growth 2023』は、さらに社員が積極的に参画するものと位置づけた。その策定では、同時に長期経営構想の見直しも行われ、その中核タスクとして、パーパスの策定が検討されたのである。

パーパス策定のベースには、2019年から始まった「地球の物語の、つづきを話そう。」のフレーズを起点としたムーブメントがあった。折しも、企業の環境保全やサステナビリティへの取り組みが、ESGやSDGsといったキーワードと共により注目を集めるようになっていた。企業に、共創、共感、社会的使命が改めて問われる時代が訪れていた。

さらに、2020年の初頭、新型コロナウイルス感染症（COVID-19）の感染拡大が全世界を襲った。この人類全体にとっての未曽有の危機は、ブランディング推

In 2020, planning began for Yokogawa's next mid-term business plan, covering the financial years 2021 to 2023. The mid-term plans are extremely important for internal branding, and the Branding Promotion Team had been actively involved in the formulation of three separate mid-term business plans in the past ten years.

In the mid-term business plan, "Transformation 2017," the corporate brand slogan "Co-innovating tomorrow" was established in line with the long-term business framework announced at the same time with the aim of aligning the efforts of its employees. To go one step further, the next mid-term business plan, "Transformation 2020," was positioned as a mid-term business plan for all employees. In order to promote understanding of the mid-term business plan and get everyone on the same page, the president and executives met in person with many employees to explain the content.

"The Accelerate Growth 2023" mid-term business plan was positioned to further encourage the active participation of employees. In the formulation of this mid-term business plan, the team simultaneously conducted a review of the long-term business framework and considered the formulation of a purpose statement as its core task.

The tagline "What's next for our planet? Let's make it smarter." introduced in 2019 served as the starting point for formulation of the Purpose Statement. Coincidentally, companies' environmental protection and sustainability initiatives as well as key expressions like environmental, social and governance (ESG) and Sustainable Development Goals (SDGs) had begun to attract attention. An era

進チームだけでなく、他の社員にとっても、横河電機の存在意義、社会的使命を見直す契機となった。これもまた、パーパス策定の後押しをすることとなった。

in which corporations began to take a closer look at co-creation, empathy, and social missions had arrived.

Furthermore, the entire world was hit by the COVID-19 pandemic at the beginning of 2020. This was an unprecedented global crisis; however, it became an opportunity for not only the Branding promotion Team but also each and every employee to rethink the purpose and social mission of Yokogawa. This also pushed the company to formulate a purpose statement.

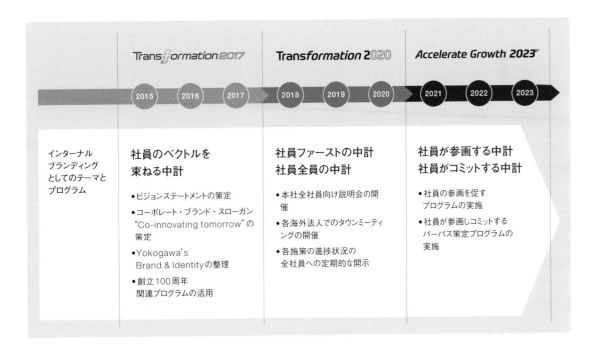

中期経営計画におけるインターナルブランディングとしてのテーマ

Internal branding themes in the mid-term business plan

「社員が参画しコミットするパーパス」の策定プロセス

The Process of Formulating the Purpose Statement with the Participation and Commitment of Employees

パーパス策定の背景となった会社の考え・思いにせよ、世界規模の課題への対処にせよ、それらを具現化するのは並大抵のことではない。そこでブランディング推進チームが考えたのが、「社員が参画しコミットするパーパス策定」というアプローチだった。

「社員との結びつきを強固にするため、社員に当事者意識を持ってもらうことは、中期経営計画、長期経営構想のテーマの一つでもありました。パーパスは社員参画でつくろうと、早いうちから決めて取り組んだのです。実は、『地球の物語の、つづきを話そう。』のキャンペーンを展開した段階で、すでに社員が気候変動などの社会課題を自分ごととして捉え、自分たちの言葉で対話して、共創につながっていく様子を目の当たりにしていたので、パーパスを社員の参画によってつくることに懸念はありませんでした」

こうして、「社員が参画しコミットするパーパス策定」の幕が切って落とされたのである。

具体的には、次の順序で作業が進められた。

①全社員に対するWebアンケートの実施

グローバル全社員を対象に、次の2つの質問を投げかけた。
質問1：10年後のYOKOGAWAをどのような会社にしたいですか。また、そうしたい理由は何ですか。
質問2：YOKOGAWAの存在意義は何だと考えますか。また、それに対してあなた自身はどのように貢献したいですか。

工夫が凝らされているのは、どちらもメインの質問に続けて「そうしたい理由は?」「どのように貢献したいか?」とサブの質問を加えていることだ。これは、当事者意識や責任意識を回答に示してほしいというブランディング推進チームの意図によるものだった。

②コメント内容の分析

質問に対して、全世界の5300人を超える社員から1万4000件を超える熱いコメントが寄せられた。
「本当に心が震えるような、自分たちはこうしたいんだという生の声が、世界中の仲間たちから集まりました」とメ

The formation of a statement that encapsulates a company's thinking on its purpose and addresses global challenges is no ordinary task. The idea of the Branding Promotion Team, therefore, was to involve employees in the formulation of this Purpose Statement.

Speaking about this, a member of the Branding Promotion Team offered the following perspective: "Giving employees a sense of ownership in order to strengthen our connection with them was one of the themes of the mid-term business plan and the long-term business framework. We decided early on that we wanted employees to participate in creating the Purpose Statement. The fact is that, at the stage when the 'What's next for our planet? Let's make it smarter.' campaign was launched, employees were already beginning to take active ownership of societal issues such as climate change. We witnessed them discussing these topics in their own words, leading to co-creation, so we were not worried about involving them in the creation of the Purpose Statement."

In this way, the Branding promotion Team kicked off the formulation of a purpose statement with the participation and commitment of employees.

Specifically, the team proceeded to work in the following order:

1) Conduct an Online Survey of All Employees

The team asked all employees worldwide the following two sets of questions:
Question 1: What do you want Yokogawa to be like 10 years from now? Why do you want Yokogawa to be as described?
Question 2: What is Yokogawa's mission to society? How do you think you can contribute to achieving the mission described?

The trick was in designing questions with the added sub-questions "Why do you want Yokogawa to be as described?" and "How do you think you can contribute to achieving the mission described?" following the main questions. The Branding Promotion Team's intention was to encourage employees to show a sense of ownership and a sense of responsibility in the answers.

ンバーは振り返る。そのいくつかを紹介しよう。

　例えば、「YOKOGAWAの存在意義は何だと考えますか?」という設問に対しては、次のような声が寄せられた。

「サステナビリティと環境貢献を目標に掲げてソリューションを提供するのはもちろんのこと、よりグリーンな未来を創造するためのコンサルティングやサービスを提供するテクノロジー企業であることと考えます」

「計測・制御・情報の技術を使って、"今"を測り、お客様の明日を守り、見えない未来を見通すことが当社の存在意義であり、お客様の明日を守ることは、地球の未来を守ることにつながると考えます」

2) Analyze the Content of the Comments

In response to the questions, 5,300 employees worldwide provided over 14,000 passionate comments.

One team member recalled, "We were deeply moved to see the comments from colleagues all over the world telling us what they wanted to do."

Here are some of the comments that were received in response to the question "What is Yokogawa's mission to society?"

"Of course, Yokogawa states that sustainability and environmental contributions are our goals and we provide solutions, but I think it should be a technology company that provides consulting and services to create a greener future"

"I think this company's purpose is to measure the present using measurement, control, and information technology, protect the tomorrows of our customers, and foresee the future that we can't see, and I think protecting the tomorrows of our customers will lead to protecting the future of the world"

Yokogawa's Purpose 策定プロセス

The process of formulating Yokogawa's Purpose

また、「10年後のYOKOGAWAをどのような会社にしたいですか?」という設問には、次のような声が寄せられた。

「当社はオートメーションと計測機器に強みを持つ会社ですが、今や製品の大半が急速にコモディティ化しています。10年後も当社が社会に貢献でき、売上と利益を維持・向上させるには、成長分野におけるバリューチェーンを進化させ、市場をリードすることが不可欠と考えます」

上記は、ほんの一例に過ぎない。コメントに目を通した社長の奈良寿氏が、「感動した」と述べたことからもうかがえるように、社員へのアンケート調査は大きな成果を収めた。

膨大なコメントは統計的に分析して回答の傾向を可視化し、キーフレーズを抽出した。同時に、ブランディング推進チームが全コメントに目を通して、統計的分析では抽出しきれないキーコメントを選び出したのである。

③社長とのラウンドテーブルの実施

国内外の20〜30代の若手・中堅社員を中心にダイバーシティーを意識し、1グループにつき8人前後のメンバーを様々な部署から集め、「キーフレーズに対してどう思うか」をテーマとして社長の奈良氏とオンラインで忌憚のない意見交換を行った。

Further, the following comment was received in response to the question "What do you want Yokogawa to be like 10 years from now?"

"This is a company whose strength is in automation and measurement devices, but most of our products are now being rapidly commoditized. To still be able to continue contributing to society and maintain and improve sales and profits in ten years, I think it is absolutely essential that we evolve the value chain in growth areas and lead the market."

The above are just some examples among many. As seen from President Nara's response that he was "deeply moved" when he looked over the comments, the employee survey was a great success.

The large number of comments were statistically analyzed, trends in the answers were visualized, and key phrases were extracted. At the same time, the Branding Promotion Team looked over all the comments and selected key comments that could not be extracted through statistical analysis.

3) Conduct Roundtable Discussions with the President

With an awareness of diversity, the Branding Promotion Team assembled groups of about eight members from various departments, centered around young and mid-level employees in their twenties and thirties both in- and outside Japan, and the team held online sessions for candid exchanges of opinion with President Nara on the topic "What do you think about the key phrases?"

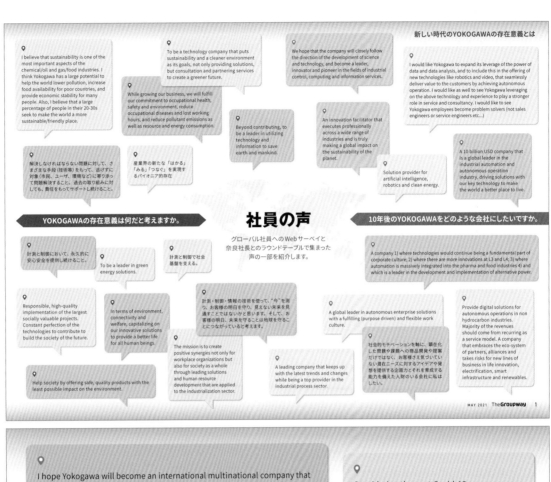

社員の声

グローバル社員へのWebサーベイと奈良社長とのラウンドテーブルで集まった声の一部を紹介します。

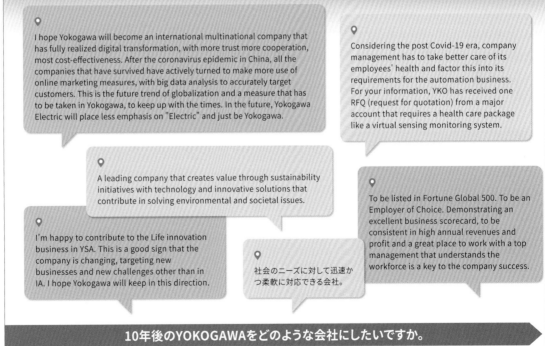

グローバル社員へのWebサーベイと奈良社長とのラウンドテーブルで集まった社員の声

Employees' ideas and suggestions collected from an online questionnaire of employees worldwide, and from roundtable discussions with President Nara

パーパス策定を通して見えてきたYOKOGAWAの姿
The Image of Yokogawa Revealed through the Formulation of the Purpose Statement

パーパスに組み込む内容としてブランディング推進チームがイメージしたのは、「企業としての提供価値は何か、何にコミットし続けるべきか、貢献領域は何か」である。さらに条件として、「SDGsに立脚しているもの」「横河電機ならではのユニークなもの」「世代、人種、言語を超えて通用するもの」「志を奮い立たせるもの」「記憶に残る長すぎないもの」「30年間鮮度を失わないもの」などが加えられた。

こうして2021年5月、中期経営計画の発表に合わせてパーパスが発表された。

Yokogawa's Purpose：
「測る力とつなぐ力で、地球の未来に責任を果たす。」

「測る」という言葉には横河電機の原点が込められており、「つなぐ」は顧客との信頼関係に加えて、同社が情報、企業、産業、価値等の結節点となることを意味している。

特筆すべきは、地球の未来に「貢献する」「支える」という表現ではなく、「責任を果たす」という強い表現が採用された点だ。これは、社員から寄せられた回答に、地球、社会貢献というワードが数多く出てきたことで、「自分たちが未来に責任を果たす」という熱いメッセージと

The Branding Promotion Team envisioned that the questions of what value Yokogawa offers as a company, what it should remain committed to, and which fields it should contribute to would be incorporated into the Purpose Statement. The team also added conditions such as being based on the SDGs, being unique to Yokogawa, not being limited to specific generations, races, and languages, being inspirational, memorable, and not too long, and not being something that would lose its freshness even after 30 years.

Thus, Yokogawa's Purpose was announced together with the mid-term business plan in May 2021.

Yokogawa's Purpose: "Utilizing our ability to measure and connect, we fulfill our responsibilities for the future of our planet."

The word "measure" embodies the origin of Yokogawa, while "connect" means that, in addition to trusted relationships with customers, the company is a hub for information, companies, industries, value, and more.

Particularly noteworthy is the adoption of the strong expression "fulfill our responsibilities" in connection with "the future of the planet," rather than expressions such as "contribute" or "support." This is because words and expressions such as "planet" and "contribute to society" appeared numerous times in the answers received from employees, so the passionate message "We fulfill our responsibil-

Yokogawa's Purpose

参画意識が感じられたためだという。まさに、社員のコミットがなければ誕生し得なかった高揚感あふれるパーパスと言えよう。

「社員の熱は相当なものであり、パーパスのベースを考えてもらうこと自体に意味があったと考えています。たとえ各自のコメントがそのままの言葉でパーパスに反映されていなくとも、自分自身で考えることには極めて価値がありました。これは必ず今後に生きることでしょう」とメンバーは語る。

同時に、10年後のYOKOGAWAのありたい姿を端的に表現したビジョンステートメントとして、「YOKOGAWAは、自律と共生によって持続的な価値を創造し、社会課題の解決をリードしていきます。」が社員の声をもとに定められた。パーパスとビジョンステートメントの策定に伴い、YOKOGAWAグループをつくり上げる要素とそれぞれが持つ役割が改めて整理され、パーパスを中心に据えたYokogawa Group Identityとしてまとめられた。

ities for the future" reflected this commitment. This is an uplifting purpose statement that would never have been created without the participation of our employees.

One team member said, "The employees are quite passionate, and we believe that the fact that we asked them to think about the basis for the Purpose Statement was meaningful in itself. Even though the exact words of every employee were not directly reflected in the Purpose Statement, there was a great deal of value in the fact that they thought about it themselves. This is something that will certainly live on in the future."

At the same time, based on the employees' comments, the team straightforwardly expressed where Yokogawa aspires to be ten years from now with the formulation of a Vision statement that reads, "Through autonomy and symbiosis, Yokogawa will create sustainable value and lead the way in solving global issues." With the formulation of the Purpose and Vision statements, various components making up the Yokogawa Group and their respective roles were reorganized and brought together in the Yokogawa Group Identity, centering on Yokogawa's Purpose.

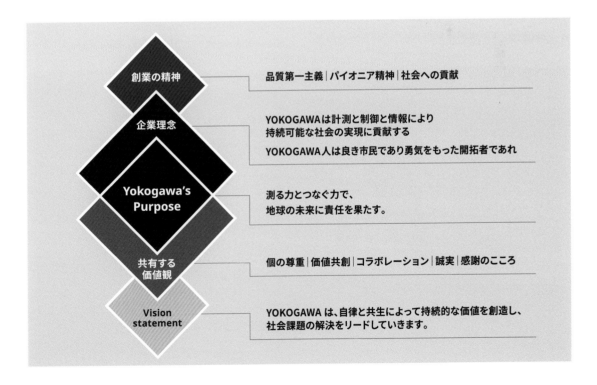

YOKOGAWAグループのアイデンティティ
Yokogawa Group Identity

新しいキー・グラフィック・エレメント「Leading Square」の発表
Announcement of a New Key Graphic Element, the Leading Square

世界を混乱に陥れたコロナ禍がまだ続く2021年、横河電機はニューノーマル（新常態）の時代を見据えた新たなブランディングに邁進する。

その活動の一つが、新しいキー・グラフィック・エレメント「Leading Square」（リーディングスクエア）の発表であった。

中期経営計画、パーパス策定のタイミングに合わせて、従来の「Brilliant Grid」に変わる新しいキー・グラフィック・エレメントとして、Leading Squareが発表された。

課題が山積するニューノーマルの時代は、企業はステークホルダーと一体となって課題解決に向かわなくてはならない。そして、次の時代をリードしていかなければならない。そのような想いを込めて、横河電機を象徴する黄色いスクエア一つにデザインを凝縮した力強い表現とした。

正方形はBrilliant Gridと同様、YOKOGAWAが評価されている信頼性、安定感、正確性を象徴している。同時にスクエアには「広場」という意味もあることから、人と人をつなぎ共創を生み出すという決意も込められている。

また、正方形のすべてを見せるのではなく、端に置いて一部を隠すことで躍動感をイメージし、社会課題解決

In 2021, while the COVID-19 pandemic that had thrown the world into confusion was still raging, Yokogawa pushed forward with new branding that focused on the era of "the new normal."

One of these activities was the announcement of a new key graphic element, the Leading Square.

At the same time as the mid-term business plan and the formulation of the Purpose Statement, the replacement of the Brilliant Grid with the Leading Square as our new key graphic element was announced.

In this era of the new normal, where challenges are piling up, Yokogawa must work together with stakeholders to focus on resolving them. And the company must then lead the way in the next era. With this in mind, the Leading Square is a powerful design in which everything is compressed into a single yellow square that symbolizes Yokogawa.

Like the Brilliant Grid, the square shape symbolizes the reliability, stability, and precision that Yokogawa is known for. At the same time, the square also brings to mind a "public square," so it also is an indication of the company's determination to connect people and generate co-creation.

In addition, not all of the square is visible. Part of the tip is hidden, making it a dynamic image, symbolizing Yokogawa's determination to take a leading role in the resolution of societal challenges.

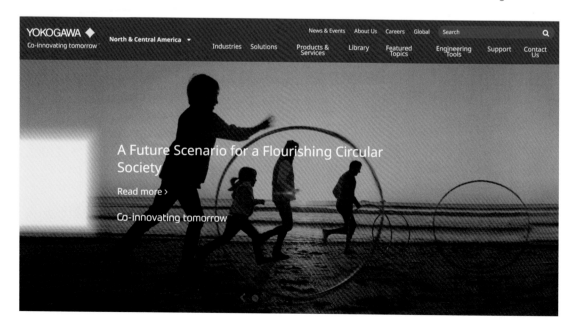

キー・グラフィック・エレメント「Leading Square」（リーディングスクエア）を使用した現在のWebサイト

Catalogs for the company, business unit concepts, and products and services layers, using the key graphic element, the Leading Square

リーディングスクエア

リーディングスクエアは、シンボルカラーであるYokogawa yellow の正方形が放射状に光輝くさまをビジュアル化したキー・グラフィック・エレメントです。

正方形（スクエア）には、製品やソリューションの精密さ、正確さ、高度さを示す とともに、人と人をつなぎ共創を生みだす「広場」という二つの意味が込められています。Co-innovating tomorrow の "Co" を表現しています。

スクエアが光り輝くさまは、YOKOGAWA がもつイノベーティブな力で社会課題の解決をリードし、未来を明るく照らしていく姿を表現しています。
Co-innovating tomorrow の "innovating tomorrow" を表現しています。

リーディングスクエア
Leading Square

活用イメージ

Leading Squareを使用したコーポレート、事業部コンセプト、製品/サービスの各レイヤーのカタログ

Catalogs for the company, business unit concepts, and products and services layers, using the Leading Square

東京国立博物館創立150年記念 特別展「国宝 東京国立博物館のすべて」の協賛広告（2022年11月11日 毎日新聞に掲載）

Advertisement in support of the "Tokyo National Museum: Its History and National Treasures" special exhibition commemorating the 150th anniversary of the Tokyo National Museum (November 11, 2022, in the Mainichi)

屋外広告（成田空港電飾看板）

Outdoor advertising (Illuminated signage at Narita Airport)

屋外広告（三鷹駅電飾看板）

Outdoor advertising (Illuminated signage at Mitaka Station)

に向けて横河電機がリードしていく意志を象徴している。

東京2020大会閉会式翌日に広告「東京の空」を出稿

東京2020オリンピック・パラリンピック競技大会（東京2020大会）は、コロナの影響を受けて1年延期され、2021年7月から9月にかけて競技のほとんどが無観客で開催された。大会の閉会式翌日に出稿したのが、「100年後の東京の空が、美しく、力強くありますように。」というコピーを添えた新聞の15段広告である。

東京の空を背景に、忍耐が求められていた社会に対して、そこで得たレジリエントな考えや未来への問いかけを広告に込めて、「地球の物語の、つづきを話そう。」というタグラインとともに世の中に発信した。

「地球の物語の、つづきを話そう。」のテレビCM

横河電機が地球環境について考え、サステナビリティに取り組んでいることを、さらに広く伝えるため、「地球の物語の、つづきを話そう。」をテーマにした新たなテレビCMを制作し、地上波、BS、屋外ビジョンなどに出稿した。

Running the "Tokyo Sky" Ad the Day After the Closing Ceremony of the Tokyo 2020 Olympics

Having been postponed for one year due to the impact of the coronavirus, the Olympic and Paralympic Games (Tokyo 2020 Games) were held without spectators from July to September 2021. Yokogawa ran a full-page ad the day after the Games' closing ceremony with the copy "In 100 Years, We Hope the Skies of Tokyo Will be Beautiful and Strong."

Against a background of the Tokyo sky, to a society that had been asked to show patience, Yokogawa filled the ad with resilient ideas derived from that patience and questions about the future, and released it to the world together with the tagline "What's next for our planet? Let's make it smarter."

The "What's Next for Our Planet?" TV Commercial

To get out the message that Yokogawa is thinking about the environment and working to achieve sustainability, a new TV commercial based on the theme "What's next for our planet? Let's make it smarter." was created and shown on terrestrial and broadcast satellite TV as well as outdoor advertising screens.

画像左：「東京の空」広告クリエイティブ（新聞）

"Tokyo Sky" ad creative (newspaper) (Left)

画像右：「地球の物語の、つづきを話そう。― 次の物語篇」TVCF

"What's next for our planet?―A new story for our future" TVCF (Right)

ブランディングをベースにした本社1階の改装プロジェクト
Brand-based Renovation Project on the First Floor of the Yokogawa Headquarters

東京・三鷹にある横河電機本社1階の改装プロジェクトに関して、総務部門からブランディング推進チームに相談があったのは2019年のことである。当初は、改装に伴うCI（コーポレートアイディンティティ）適用の確認レベルを想定していたが、コロナ禍が契機となり空間の役割をゼロから考え直すことになった。

ブランディング推進チームが考えたコンセプトは、単なるショールームにするのではなく、ブランドを体験する場にするというものだった。

「コロナ禍でお客様と対面で会う機会が減っていましたが、いつかは必ず以前のように世界中のお客様が本社にいらっしゃるようになります。そのときに、どういう空間でお客様と関わるべきなのか。様々な環境変化や市場の要求変化など、どのような状況においても横河電機が信頼に足るパートナーであることを確信していただける空間を目指しました」とメンバーは語る。

3つに分かれたゾーンには、それぞれプレゼンテーションスペース、対話スペース、コミュニケーションラウンジが設けられ、文字通りYOKOGAWAのブランドを体験できる場として、様々なイベントが開催されている。

In 2019, the Branding Promotion Team was consulted by the General Affairs Department about a project to renovate the first floor of Yokogawa's headquarters in Mitaka, Tokyo. Initially, the team anticipated that they would simply need to confirm the applicability of the corporate identity (CI) in conjunction with this renovation, but the COVID-19 pandemic triggered a complete rethinking of the role played by this space.

Rather than this being a simple showroom, the concept developed by the Branding Promotion Team was for it to be a place where one could experience the brand.

One team member said, "The COVID-19 pandemic has reduced opportunities to meet with customers face-to-face; but someday, customers from around the world will surely start visiting our headquarters again. At that time, what kind of space should we have for engaging with them? Our goal was a space that would offer customers assurance that Yokogawa remains a trustworthy partner in every circumstance, whether that involves responding to challenges such as a changing climate or a rapidly changing market."

The team divided the space into three zones—a presentation space, a discussion space, and a communication lounge. In this place where the Yokogawa brand can literally be experienced, a variety of events are held.

ブランド体験エリア「Brand experience zone」

Brand experience zone

ブランディングの新たな展開
New Launch of the Brand

　2013年に始まった横河電機のブランド再構築のプロジェクトは、全世界の社員や顧客を巻き込みながら2023年現在までの10年間、着実に進んできた。

　経営陣とブランディング推進チームの情熱、そしてブランドアンバサダーの自覚をもって参画してきた社員の熱意によって、新しいYOKOGAWAブランドは市場や社会に浸透したと言ってよいだろう。今や、計測・制御事業のメーカーとしてではなく、時代の最先端に立って社会課題を解決し、地球環境に貢献するソリューション提供企業として、横河電機は国際的に認識されつつある。

　そのことを証明するように、2019年1月、サステナビリティインデックスの一つとして広く知られている「世界で最も持続可能な100社（Global 100 Most Sustainable Corporations in the World Index）」に選出された。2022年には、フランスのEcoVadis（エコバディス）によるサステナビリティ調査において、評価対象企業の上位5%以内の企業が認定される「ゴールド」の評価を獲得している。

　だが、社会環境が激変し、世界情勢が混迷を極める中、ブランディングは立ち止まることなく継続しなければならない。ブランディング推進チームは、2023年3月に、「地球の物語の、つづきを話そう。」の次の展開として、「物語のつづき。夢のはじまり。」をキャッチコピーとした広告を出稿。未来世代が美しい夢を見続けられるように、「地球の物語」を紡いでいくことを広く伝えた。「今、目指しているのは、パーパスを起点としたブランデ

Yokogawa's brand reestablishment project began in 2013 and has involved employees and customers all over the world, and as of 2023 has been making steady progress for ten years.

With the enthusiasm of the management team and the Branding Promotion Team and the zeal of employees who have participated with the awareness of being brand ambassadors, the new Yokogawa brand has inarguably gained recognition in the market and with the general public. Yokogawa is now internationally recognized not just as a manufacturer engage in the measurement and control business, but as a company that is on the cutting edge of the current era, helping to come up with solutions to issues such as a changing climate that concern everyone in our society.

As proof of this, in January 2019 Yokogawa made it onto the Global 100 Most Sustainable Corporations in the World Index, which is one of the world's leading sustainability indices. In 2022, the company acquired a Gold rating, recognizing it as being in the top five percent of companies assessed in a sustainability survey conducted by France's EcoVadis.

However, in a rapidly changing society and with world affairs currently in great confusion, Yokogawa must continue its branding. In March 2023, the Branding Promotion Team launched the follow up to "What's next for our planet? Let's make it smarter." by running ads with the catchphrase "The next chapter. It begins with a dream." With these new ads, Yokogawa is continuing to share far and wide stories for our planet in which future generations can continue to dream beautiful dreams.

ィング、社員が参画するブランディングです。グローバルの社員全員が社会的使命や時代の要請を意識しながら、自らが描き出したパーパスを体現し、日々の業務を通じてブランドアンバサダーやスポークスパーソンとしての役割を担っていくことです」とブランディング推進チームのメンバーは語る。

　横河電機のブランディングの成功は、誰もが納得し、共感できる物語を紡ぐことで、社員をはじめとするステークホルダーを巻き込んだことが大きな要因と言えよう。横河電機はブランディングを通じて、Co-innovating tomorrow、「地球の物語の、つづきを話そう。」、Yokogawa's Purposeという、将来を切り開くための3つの大きな指針を手に入れた。これからも、将来に向けた考えや背景にある思いを、空想の物語ではなく、確信と信念を込めたブランドのストーリーとして語っていくための取り組みが続けられていくに違いない。

A Branding Promotion Team member said, "What we're aiming for now is a brand with the Purpose Statement as the starting point, and a brand in which our employees participate. While all of Yokogawa's employees worldwide are aware of our social missions and the demands of these times, they will personify the Purpose Statement they imagined themselves and, through their tasks each day on the job, take on roles as brand ambassadors and spokespersons."

The involvement of our employees and other stakeholders in creating a story that everyone can understand and identify with has been a major factor in the success of Yokogawa's branding. Through this activity, Yokogawa has come up with three major guiding statements that point the way to a brighter future: "Co-innovating tomorrow," "What's next for our planet? Let's make it smarter." and Yokogawa's Purpose. Going forward, there is no doubt that Yokogawa will continue to share its thoughts on the future, not as a fantasy but as a brand story that is full of conviction and belief.

一橋ビジネススクール 国際企業戦略専攻 客員教授
名和 高司 氏
Takashi Nawa
Visiting Professor
Hitotsubashi University Business School
School of International Corporate Strategy

×

横河電機株式会社 常務執行役員 マーケティング本部 本部長
阿部 剛士 氏
Tsuyoshi Abe
Senior Vice President
Head of Marketing Headquarters
Yokogawa Electric Corporation

無形資産としてのブランド価値
Brand Value as Intangible Assets

横河電機の本社1階にあるブランド体験エリア「Brand experience zone」に、一橋ビジネススクール客員教授の名和高司氏を迎え、同社マーケティング本部長の阿部剛士氏とともに、無形資産、そしてそこから派生する企業のブランド価値について語り合った。（以下、敬称略）

——信頼やブランド力といった無形資産が、企業価値を測る基準の一つとして注目されつつあります。無形資産とブランド価値には、どういう関係性があるのでしょうか。

名和 無形資産は非財務資産ではなく、"未財務資産"と呼ぶべきだと考えています。単なる非財務資産だと企業価値には結びつきにくいのですが、無形資産はやがて企業価値に結びつく可能性が高いので"未財務"と呼ぶのがふさわしいのです。設備やマネーなどの有形資産を有効活用させるのが無形資産であり、それは企業の将来価値を示すPER（株価収益率）のドライバーでもあります。

　無形資産は大きく"組織資産""顧客資産""人財資産"の3種類があり、組織資産は企業が持つ文化や価値観などのコアの部分、顧客資産は企業のファンであるお客様を示しています。そして、この2つを伸ばす大事な要素が、企業の中の人である人財資産です。

　しかし、人がパワーを持って集まるだけでは、まとまりのない集団になってしまいます。そこで、人の思いを一定方向に駆動したり収束したりするのがブランドだと私は思っています。

　ブランドには、顧客や市場など外部からどう受け取られているかというアウターと、内部の社員がそれをどう自覚しているかというインナーの2つの側面があります。そして、社員自身が何を達成したいのかを自覚し、それ

We invited Takashi Nawa, a visiting professor at Hitotsubashi University Business School, to discuss the topics of intangible assets and corporate brand value with Tsuyoshi Abe, a Yokogawa senior vice president who heads the company's Marketing Headquarters. This discussion took place in the Brand experience zone on the first floor of the main building at the Yokogawa corporate headquarters.

Intangible assets such as trust and brand strength are receiving increasing emphasis as yardsticks for the evaluation of corporate value. What do you think about the relationship between intangible assets and brand value?

Professor Nawa: I think we should consider intangible assets not as non-financial assets, but as "potential" financial assets. Non-financial assets tend not to be connected to corporate value; however, intangible assets have the potential to become corporate value. Therefore, it is appropriate for us to call them "potential" financial assets. Intangible assets make use of tangible assets such as equipment and money. For this reason, intangible assets are the driver for increasing the price-earnings ratio (PER), which shows the future value of the corporation.

Intangible assets include three components: organizational assets, customer-related assets, and human assets. Organizational assets are the core of corporate culture and sense of value, and customer-related assets are customers as fans of the corporation. The important component that increases these two types of assets is human assets.

However, simply gathering people with individual strengths does not develop a well-balanced group. Therefore, brand is important in driving and converging individual intentions in one direction.

Brand has two aspects. One is the external aspect,

写真左：阿部 剛士 氏
写真右：名和 高司 氏

Mr. Tsuyoshi Abe (Left)
Mr. Takashi Nawa (Right)

が仲間や顧客の胸の中に投影されて刻まれていくとブランド価値になると私は考えています。

——ブランド価値を含めた無形資産を最大限に活用するには、どういった戦略や取り組みが必要だと考えますか。

名和　エンゲージメントとアルゴリズムがキーワードだと思っています。エンゲージメントとは、「こういう世界をつくりたい」「志に共感する」という共創や共感をもとに、様々なステークホルダーを引き付ける力のことです。

　加えて、それを実現するためのプロセスをアルゴリズムとして持つ必要があります。このアルゴリズムは、その企業特有の価値のつくり方という意味で、価値創造方程式と呼んでいます。分かりやすく言うと、あるインプットがあったときに、それをうまく処理してアウトプットを生み出し、さらにそれを社会に実装してアウトカムにするという一連の流れの中で、その会社らしい付加価値を生む仕組みが、ここでいうアルゴリズムです。

　実は、日本企業の弱点がここにあると私は思っています。日本企業には匠の技術があるので、何でもやれば

which is the customers' and markets' perspectives of the brand. The other is the internal aspect, which is the employees' perspective of the brand. Ensuring that employees are aware of what needs to be achieved, and that they share this awareness with other workers and customers leads to the establishment of brand value.

What strategies and approaches are needed to make maximum use of brand value and other intangible assets?

Professor Nawa: I think engagement and an algorithm are the keys. Engagement is the power to attract a wide range of stakeholders based on a shared spirit of co-creation such as the ideal society that they would like to create.

In addition, it is also necessary to develop a process to achieve such intentions as an algorithm. We call the algorithm the "value creation formula," which means a method of creating unique corporate value. When receiving input, we process that input to create output, and implement it into society to create an outcome. The structure that produces unique corporate added value in the flow of such a process is the algorithm that I mean here.

Profile_Takashi Nawa

東京大学法学部卒、ハーバード・ビジネス・スクール修士（ベーカースカラー授与）。三菱商事の機械部門（東京、ニューヨーク）に約10年間勤務。2010年までマッキンゼーのディレクターとして約20年間、コンサルティングに従事。自動車・製造業分野におけるアジア地域ヘッド、ハイテク・通信分野における日本支社ヘッドを歴任。日本、アジア、米国などを舞台に多様な業界において、次世代成長戦略、全社構造改革などのプロジェクトに幅広く従事。2010年6月より、一橋大学大学院国際企業戦略研究科教授に就任。

Takashi Nawa received his MBA from Harvard Business School (as a Baker Scholar), and his Bachelor of Arts in Law and Political Science from the University of Tokyo. He worked at Mitsubishi Corporation Machinery Division (Tokyo, New York) for about 10 years. He worked at McKinsey & Company as a Director engaged in consulting for about 20 years until 2010. He served in the automobile and manufacturing sectors as a Practice Leader for the Asia-Pacific Region, and the high-tech and telecommunication sectors as Practice Leader for the Japan Office. In Japan, throughout Asia, and in the United States, he engaged in a broad range of projects such as next-generation growth strategies and company-wide structural reformation in different industries. Since June 2010, he has served as a professor at Hitotsubashi University School of International Corporate Strategy.

できてしまうのですが、どうすればよいのかを言語化できないのです。

　私は以前から、「たくみ（匠）のしくみ（仕組）化」と提唱しています。自分たちならではのプロセスを言語化できれば、その会社らしい再現性ができて、企業価値が高まることは間違いありません。

阿部　おっしゃる通り、エンゲージメントのベクトルが合うと、10人が30人の働きをすることを実感しています。ただ、多くの日本企業ではそのベースとすべきビジョンやパーパスが冗長なために、社員が理解できず、ベクトルを合わせにくいのです。人間の脳は分かりやすいものを好みます。シンプルで整合性のあるメッセージにすることが大切です。

　また、エンゲージメントとアルゴリズムをもとにブランド価値を維持していく際に重要なのは、外部環境変化に対応したブランド価値マネジメント戦略だと考えています。私は、ブランドとは社内外に対する"コミットメント"と位置づけています。そのためブランドに一貫性がある

In fact, this is a weak point of Japanese companies. They have excellent technologies, which allow them to succeed at anything; however, they are not able to verbalize them.

I have suggested that we need to formulate craftsmanship into algorithms. If we can verbalize our unique process, I am sure that individual companies can improve reproducibility, which will definitely increase corporate value.

Dr. Abe: If the engagement vectors are aligned, then ten people can achieve the results of thirty. However, the vision and purpose statements that serve as the foundation for engagement in Japanese companies are often overly verbose, and this gets in the way of employee understanding and the alignment of vectors. People have a natural preference for things being easy to understand, so it is important to create simple and consistent messages.

Furthermore, with the use of engagement and algorithms to maintain brand value, it is important to have a brand value management strategy that can respond to

ことは基本ですが、これだけ技術の進歩や社会環境の変化があると、一貫性を維持しつつブランドをメンテナンスしていくこともまた欠かせません。

名和　一貫性を持ちながら外部環境に対応するのは二律背反で難しく、きちんとした軸を持っていないと単に流されるだけになってしまいます。軸を持ちながら進化させるには、バスケットボールのピボットをイメージするとよいでしょう。ピボットというテクニックは、軸足さえ動かさなければ、もう一方の足は360度動かすことができます。しかし、軸足を動かすと見当違いの方向に進んでしまいますし、だからといって両足とも動かさなくては何も進歩がなくなってしまうのです。

——ブランド価値を維持するには人財資産がポイントというお話がありましたが、どのようにして社員に無形資産やブランド価値を理解してもらい、それをどう共有していけばよいのでしょうか。

名和　インナーブランディングの観点からすれば、トップが強引に推し進めるのではなく、ビジョンやパーパスを社員に"自分事化"してもらうことが大切です。

それには、コーポレートのビジョンやパーパスを、まず部門事や組織事にすること、さらに一般社員の自分事に落とし込んでいくという、2段階のプロセスが必要となってきます。「すごくいいことを会社は言っているけれども、では私たちはどうすればいいの?」と社員が戸惑わないように、それぞれの現場でビジョンやパーパスを落とし込むブランドアンバサダーが必要になるのですが、それには現場に近いミドルの人がキーになると思います。

同時に、アウターブランディングを通じて顧客や社会から評価を受けると、社員は自分たちがその一翼を担っていると実感でき、ワクワク感につながることでしょう。BtoB企業の経営層の中には、「お客様は自分たちをよく知っているから、ブランディングは必要ない」という人がいますが、そうではありません。

阿部　同感です。「BtoBとBtoCは、マーケティングの観点からどう違うのか」とよく聞かれますが、最も重要なのは価値基準の優先度だと思っています。BtoC

changes in the external environment. I consider a brand to be the commitment that a company makes to people both in- and outside the organization. It is therefore essential to maintain consistency with the brand. However, one must also make sure that the brand keeps pace with technological advances and societal change.

Professor Nawa: Corresponding to external environmental changes while maintaining consistency sets up a contradiction that makes aligning the two difficult to achieve without a stable axis. Making progress while maintaining an axis is like pivoting in basketball. Pivoting allows one foot to rotate 360 degrees while the other remains stationary. However, when you move the pivoting foot, you may move in an unintended direction. But you need to try because if you don't move both feet, you make no progress.

You said human capital is key to maintaining brand value. How can one ensure that employees have a shared understanding of intangible assets and brand value?

Professor Nawa: From the viewpoint of inner branding, top management should not be the driver. It is important that the employees share the visions and purposes among themselves.

In order to do so, a two-step process is necessary. The first step is to ensure that individual divisions throughout the company understand corporate visions and purposes. The second step is to then ensure that each employee understands them. To avoid confusion among employees, who might think, "The company has great ideas, but we don't know how to translate them into action," we need to have brand ambassadors who help individual employees fully understand the corporate visions and purposes. Otherwise, they may lose their focus. And I think mid-career employees who fully understand on-site employees will be the key to this.

At the same time, maintaining a good reputation among customers and in society through outer branding motivates employees by assuring them that they are playing an important part in outer branding. Some top management at BtoB companies think that branding is unnecessary because customers know them well. But this is incorrect.

の場合はモノやサービスの品質自体が基準になります
が、BtoBの場合は何よりも企業としての信頼が優先さ
れます。そして、信頼できる会社かどうかはブランドなく
しては語れません。そのことに日本のBtoB企業は早く
気づくべきでしょう。

――パーパスは無形資産の一つとして重要なものです
が、パーパス策定が経営にプラスになっている企業とな
っていない企業の差はどこにあるのでしょうか。

名和 良いパーパスの例として私が挙げるのは、「ワク
ワク」「ならでは」「できる!」の3条件がそろっているもの
です。パーパスを聞くとみんなが「ワクワク」した気分
になり、心のエネルギーになるのが第一条件です。そ
れ以上に大事なのは「ならでは」で、他の会社では言
えない、その会社ならではの要素が含まれているかどう
かがポイントです。その2つがあれば、社員や顧客が「で
きる!」と感じ、実行に移していけるパーパスになります。

社員が共感できるパーパスにするには、社員を巻き
込んだパーパスづくりのプロセスが大事だと思います。
社長の鶴の一声で決めたパーパスは、とりあえずつくっ
て飾っておくだけの"飾り事パーパス"になりがちです。

その点、横河電機のパーパスは、社員のみなさんが
参画したワクワク感たっぷりのものであり、しかも横河
電機ならではの「測る力」と「つなぐ力」というコアコン
ピタンスを含む素晴らしいものだと思います。あとは組
織事、自分事へどう落とし込み、企業価値につないで
いくかが大切だと思います。

――激変する社会環境の中で、今後は無形資産やブ
ランディングはどのような方向に進んでいくのでしょうか。

名和 先ほども触れましたが、日本のBtoB企業に必
要なのは「たくみ」を「しくみ」にすることだと考えていま
す。一般的に日本の企業は顧客の要望から逃げないで
うまくやるのは得意ですが、技術が属人化してしまい、
再現性が欠けてしまう問題があります。「たくみ」の技を
「しくみ」として標準化できれば、横展開によってビジネ
スのスケールアップやスピードアップが可能になります。

阿部 確かに、横河電機には伝統的にエンジニアリン

Dr. Abe: I agree with you. I am often asked what the difference between BtoB and BtoC is from a marketing perspective. I think the most important difference is in what they prioritize. BtoC prioritizes the quality of products and services, while BtoB places the top priority on company credibility. And it is impossible to assess corporate credibility without there being a brand. Japanese BtoB companies must realize this fact soon.

A purpose is an important intangible asset. What is the difference between a purpose that makes a positive contribution to a company's business, and one that does not?

Professor Nawa: Good purposes, I believe, are those that satisfy three conditions—excitement, uniqueness, and capability. First off, purposes must excite employees and create psychological energy. Most important is the uniqueness of the company. If the company has at least two of these elements, employees and customers can trust its capability, which leads to actual implementation.

To determine purposes that all employees embrace, it is important to involve employees in their determination. Purposes chosen solely by the president tend to be meaningless, mere decorations.

The purposes that Yokogawa establishes create excitement through the full participation of employees. They also include the core competencies as well as the measuring and connecting capabilities of the company. It is then important to consider how to share purposes throughout the company, and how individual employees can implement purposes to improve corporate value.

As the world around us experiences rapid change, what trends can be anticipated with intangible assets and branding?

Professor Nawa: As I mentioned before, I think Japanese BtoB companies need to formulate craftsmanship into algorithms. Generally, Japanese companies are good at dealing with difficult requests from customers; however, craftsmanship becomes individualized and less reproducible. If we could formulate craftsmanship into algorithms, it would be possible to increase the scale and speed of business through horizontal expansion.

Profile_Tsuyoshi Abe

1985年、インテルジャパン（現インテル）に入社。2005年、同社マーケティング本部長就任。2007年、芝浦工業大学専門職大学院 技術経営/MOT修了。2009年、同大学地域環境システム専攻博士課程修了。2011年、同社取締役副社長兼技術開発・製造技術本部長に就任。2016年、横河電機に入社、現在に至る。

Tsuyoshi Abe joined Intel Japan (current name Intel K.K.) in 1985. In 2005, he became the Senior Vice President of that company's Marketing Headquarters. In 2007, he received a Management of Technology (MOT) degree from the Shibaura Institute of Technology (SIT) Graduate School, and in 2009 earned a doctorate in Regional Environment Systems from SIT. In 2011, he assumed the position of Senior Vice President and Director of the Technology and Manufacturing Group at Intel. In 2016, he joined Yokogawa Electric Corporation.

グ・テクノロジーという強みがあるのですが、残念ながらほぼ暗黙知といってよい世界です。それを形式知化して、最適なモデルを構築できれば、再現性が高まって企業の信頼にもつながると感じています。

　日本のBtoB企業の多くは、これまでは目に見えないものになかなか価値を見いだそうとしませんでした。しかし、技術革新によってブランドが定量化、可視化できるようになってきた現在、無形資産も資産の一つとして、企業の経営者は腹を据えて認識すべきときが来ていると感じています。

名和　ブランドは定性的な見方に流れがちですが、それを横河電機らしく、「測る力」を発揮して見える化することが大切です。横河電機には、ぜひ日本のBtoB企業をリードするブランディングのお手本になっていただきたいと願っています。

Dr. Abe: Indeed, Yokogawa has traditionally been strong in the area of engineering technology, but this, unfortunately, is almost all tacit knowledge. Changing it into explicit knowledge and establishing optimal models would increase reproducibility, and this in turn would lead to the establishment of corporate credibility.

Many Japanese BtoB companies have not tried to discover value in items that are less tangible. However, now that technological innovations are making it easier to quantify and visualize brand value, I feel that the senior executives of these companies should recognize that intangibles can also be assets.

Professor Nawa: Brand tends to be viewed qualitatively. However, it is important for Yokogawa to exercise its ability to measure and visualize brand value. My hope is that Yokogawa will become a good branding model that will lead BtoB companies in Japan.

ここまで、横河電機が2000年代以降に取り組んできた企業アイデンティティの再定義と、その基盤に立脚したブランド再構築の軌跡をたどってきた。グローバルなBtoBビジネスを展開する同社らしい形で、世界中の社員を巻き込む全社的な取り組みは現在も続いている。これらの活動の中心に立ったのは、ブランディング推進チームの4人のメンバーである。この4人がどのような想いを抱き、何をモチベーションにしてこのプロジェクトに取り組んだのか、それはメンバーから寄せられた、以下の文章を通じて明らかになるだろう。

The preceding chapters have traced how Yokogawa has redefined its corporate identity since the 2000s and reestablished its brand based on this foundation. Group-wide efforts involving employees from around the world continue even today in forms that are consistent with Yokogawa's global BtoB business operations. The central players in these activities are the four members of the Branding Promotion Team. The following sections were submitted by these four people and describe the aspirations they had and what motivated them as they conducted the rebranding project.

次の100年、200年をつくっていくために

Setting the Stage for the Next 100 to 200 Years

横河電機は100年以上の歴史を有していますが、本書の内容は主に2013〜2023年現在までのブランディングをまとめたものです。

会社の次の100年、200年をつくっていくためには、事業の発展に寄与する戦略的なブランディングをさらに強化しなくてはならないと考えています。国内外におけるYOKOGAWAのブランド認知はまだまだ十分とは言えません。ここまで培ってきたことを礎とするだけではなく、さらなる飛躍のための一歩と考え、これからもまい進していく所存です。

BtoB企業のブランディングには他にも様々な手法や考え方があります。当書籍を一つの事例としてお読みいただき、私どもの事例をきっかけに、より良い解をご指導いただける機会、共創の機会につながれば、大変うれしく思います。

While Yokogawa has over 100 years of history, this book mainly summarizes the branding activities undertaken between 2013 and 2023.

To set the stage for the company's next 100 to 200 years, we need to further strengthen our strategic branding for the development of our business. Brand awareness for Yokogawa both in- and outside Japan is still not where it should be. We intend to use what we have built so far not only as a foundation but also as a springboard in our continuing efforts to achieve gains.

Many other tactics and approaches can also be applied to BtoB corporate branding. I would be delighted if this brand book is read like a case study and provides guidance that will lead to even better solutions and opportunities to engage in co-creation.

社員なくして、大きな木は育たない
A Large Tree Cannot Grow without Employees

エイブラハム・リンカーンが遺した名言があります。

Character is like a tree
and reputation like its shadow.
The shadow is what we think of it.
The tree is the real thing.
人格は木のようなものであり、
評判はその影のようなものである。
影とは、我々が人の性格をどう思うかということであり、
木こそが本物である。

業務を行っている中で、かつて教えられたこの言葉を常に意識しています。

ブランドに関連する活動は、得てして評判をつくるためのものと思われがちですが、真の目的はコーポレートキャラクターである "木" をしっかりと育てること、その木をつくり上げている社員にその自覚を持ち活躍してもらうことが大切だと考えています。

2015年にブランディング推進チームのメンバーになった当時、横河電機の社員は会社に誇りを持ち、会社に愛情を持っていると実感しました。当社がお客様に評価されている「信頼できる」「顧客のニーズへの対応に熱心」「誠実である」というイメージは社員一人ひとりから感じられました。一方で、対外的には事業ベース、個人ベースでの表現が散在し、まとまった力として発揮できていない面も見受けられました。会社としてもっと効果的に打ち出していく方法さえ提供できれば、もう一歩先のステージに進める。そんな気がしていました。

私は、本社が明確な方針を示し、指示を出し、その上で各国市場に即した活動を許容すれば、グループ企業として活動を進めやすいことを経験していました。そのため、詳細なガイドラインを作り、見本となるコンテンツを制作する一方で、海外拠点のマーケティングメンバーの要望には可能な範囲で柔軟に対応しました。今では、グループとしての一体感が醸成し始めたと感じています。本社の毅然たる姿勢と真摯な対応は、グローバルでのブランディングにとってはこの先も重要なことでしょう。

そして、改めてコミュニケーションを業務として担当す

There is a famous saying by Abraham Lincoln.

Character is like a tree and reputation like its shadow.
The shadow is what we think of it. The tree is the real thing.

I've heard this saying before, and I keep it at the top of my mind when I'm engaged in my work.

Reputation building tends to be thought of as the purpose of brand-related activities, but the true purpose is cultivating the tree, i.e., corporate character, and it's vital that the employees who are growing the tree have this awareness as they perform their work.

After joining the Branding Promotion Team in 2015, I came to see how Yokogawa's employees took pride in their company and felt loyal to it. It seemed that the employees had the same image of Yokogawa that customers praised, namely its dependability, drive to meet customers' needs, and integrity. Conversely, it also seemed that expressions used externally for this image varied from business to business and person to person, and that Yokogawa's image was not being deployed as a collective strength. If we could just offer the company a more effective means of expressing its image, it could move to the next stage. This is what I was thinking.

I experienced how much easier it is to conduct brand activities as a company in a global group of companies when the headquarters sets forth a clear policy, gives instructions, and then makes allowances for activities that are localized for each national market. For this reason, while we created detailed guidelines and produced model content, requests from marketing staff at our locations outside Japan were flexibly accommodated as much as possible. It feels like we are now starting to cultivate a sense of unity throughout the Yokogawa Group. The headquarters' steadfast character and integrity will continue to be important in future global branding.

As someone who works in communications, I have through this experience gained a keener appreciation of how our activities have connected

るものとして、人とのつながり、人を支え、支えられ活動をしてきたことを実感しました。人なくして、大きな木は育ちません。

私たちのこの10年の活動はまだまだ礎を作りかけたところ。これから先は、次の世代が育てていってくれると信じています。

私たちの仕事は、卓越した企業、ひいては卓越したブランドをつくることです。

people, supported them, and received their support. A large tree cannot grow without people.

Our activities over these ten years have only just created the foundation. Looking forward, I am confident that the next generation will develop this foundation.

Our work is to create a great company, and in turn, create a great brand.

企業文化変革装置としてのブランディング
Branding as an Apparatus for Transforming Company Culture

私が当社に入社したのは、「横河北辰電機」が「横河電機」に社名変更する前です。当時はプロセス制御事業に国内の大手重電ブランドが相次いで参入し、当社はその生き残りをかけて社名変更と新しいシンボルの開発を進めていました。

現在のシンボルは、「新しい時代にチャレンジするための挑戦であり、日本の横河から世界のYOKOGAWAへと大きく成長するための道標」として開発されたものです。

企業は環境の変化に対応し、社会に求められる存在でなければ持続的に企業価値を向上させることはできません。そのためには自社の「企業文化」も時代と共に進化させる必要があり、このとき、ブランディングが一種の「企業文化変革装置」の役割を果たします。顧客、社員、株主など、異なった視点を持つステークホルダーを一つにまとめ、新しい方向に導きます。

組織が大きくなればなるほど個人の存在が相対的に小さくなり、企業の全体像や役割が見えにくくなります。そんなとき、ブランディングで各自がYOKOGAWAを代表する重要なメンバーであることを認識させ、正しい振る舞いに導くのです。当社はヒューレット・パッカード、ゼネラル・エレクトリック、クレイ・リサーチといった企業と合弁事業を行うなど、時代の要請に合わせ新しい能力を獲得してきました。その意味で、この会社には「新しい力を活用する」資質があると言えます。

I started working at Yokogawa Electric when it was still Yokogawa Hokushin Electric, before it changed its name. At the time, many leading Japanese heavy electric equipment brands were entering the process control business, and Yokogawa was in the middle of the process of changing its name and developing a new corporate symbol as part of its efforts to survive this competition.

The current corporate symbol was developed to serve as a guidepost as we took on the challenges of a new era and sought to achieve significant growth by transforming from a Japanese Yokogawa to a global Yokogawa.

For a company to sustainably increase its corporate value, it must be able to adapt to changes in the external environment and continue to be needed by society. For this reason, Yokogawa's company culture also needs to evolve with the times, and branding is serving as one apparatus for transforming the company culture at this time. Branding can unite stakeholders such as customers, employees, and shareholders that have different viewpoints, and guide them in a new direction.

As an organization gets bigger, the relative weight of the individuals in it gets smaller, and it becomes more difficult for each person to form a complete picture of the company and its roles. Branding then reminds individuals that they are important representatives of Yokogawa and guides them to take the right actions. Over the years, Yokogawa has acquired new capabilities that have been

しかし、それらを内部から「企業文化」という視点で問い直したとき、全く違う姿が見えてきます。高度成長期、世界に製品を届けていたものの、中長期的に将来のビジネスを発展・成長させるための「企業文化」が全く醸成されていない。少なくとも私の目にはそう映っていました。「このままではいけない。そのうち大きなブランド毀損を生じさせるのではないか」と切実に感じていました。

そんな思いを持ちながら創立100周年に向けて準備を進めていたところ、あるグローバル企業から移ってきたメンバーがマーケティング部門に加わりました。その一人が、現在のブランディング推進チームのリーダーです。当時、経営管理本部でコーポレートアイデンティティの管理をしていた私は、ある日この気持ちをぶつけ、こうお願いしたのです。

「当社は海外売上高比率が約7割の会社ですが、それらビジネスを支えるに見合った企業文化が醸成されていません。ブランド管理の視点から見た場合、東京・三鷹の中小企業がそのまま大きくなった状態で、大変危機感を持っています。ぜひ、有名なグローバル企業に従事された経験と知見を活かしていただきたい。サポートいただきたい」

そのときのことは、今でも情景を描けるほど鮮明に覚えています。創立100周年事業の準備そのものが、まさしく我々にとって"Co-innovating tomorrow"のスタートだったのです。それからというもの、新しい仲間と共に、制御事業という枠組を超えて明日のためのブランディングに挑戦し続けてきました。

当社のような純粋なBtoB企業のブランディングが書籍として出版される機会は少ないと思います。その意味で、この書籍が少しでもBtoB企業の皆様の参考となりお役に立つことを切に願っています。

suited to the needs of the times, by such means as engaging in joint ventures with companies like Hewlett Packard, General Electric, and Cray Research. In this sense, you could say that Yokogawa has the capacity to leverage new capabilities.

But when you reframe these activities from the internal perspective of the company culture, a different picture emerges. While the company delivered products worldwide during the period of rapid economic growth in Japan, it did not cultivate a company culture that would allow it to grow and develop its business over the medium- to long-term. At least that's how I viewed it. I urgently felt that we could not continue as we were, and that if we did, we would end up significantly damaging our brand.

With these thoughts in mind, I was working on the preparations for our 100th anniversary when a new person joined the Marketing Headquarters after leaving a global company. This person is the leader of our current Branding Promotion Team. At the time, I was in the Corporate Administration Headquarters and was responsible for managing our corporate identity. One day I could no longer contain this feeling I had and made the following request:

"We're a company with global sales that account for approximately 70% of our business, but we have not cultivated a commensurate company culture to support this business. From the position of brand management, Yokogawa seems more like a small business enterprise from Mitaka, Tokyo, that just happened to become a big company, and I see this as a big crisis. I would like to ask you to leverage the experience and expertise you have from working for a renowned global company and support us."

I still remember this so vividly today that I can picture the scene in my mind. The preparations for the 100th Anniversary Project were for us the true start of "Co-innovating tomorrow." Since then, we have continued to pursue branding together with new colleagues to create a future that goes beyond the industrial automation and control business.

I do not think many pure BtoB companies like ours have had the opportunity to publish a report in book form on their branding activities. Accordingly, I very much hope that this book serves as a useful guide for other BtoB companies.

『物語のつづき。 夢のはじまり。』

The next chapter. It begins with a dream.

ちょうど10年前、横河電機に転職してきたときに感じたことは、この会社には社会に大きな影響力を持ったお客様が世界中に存在し、競争力のあるテクノロジーがあり、お客様と現場を大切にする多くの社員がいるということでした。一方で、ブランドの視点では、そうしたアセットが会社のブランド価値として集約されておらず、統合的なコミュケーションも不在で、ブランド露出のボリュームにおいては、当社のビジネス規模、上述のアセットの価値と比較すると、決定的と言えるほど小さいと感じました。

ブランド露出のための予算を増やすことは、一朝一夕でかなうものではありません。だとしたら、何から着手できるのか……。まずはブランドをシステムとして考え、ばらばらな方向に向いているアセットをブランド価値として束ね、社員をブランドアンバサダーとして動機付けることではないか。

ここから私たちのブランド再構築プロジェクトは始まりました。

求心力のある御旗を掲げ、ブランドの体現者である社員をエンゲージし、社員自らが自分たちの社会的使命を考え、発信する……。そうした仕組みづくりを10年かけて行ってきました。本書では、その仕組みづくりの過程を紹介しています。

もう一つ本書で紹介しているのは、仕組みづくりと並行しながら、私たちブランディング推進チームが社内外に提示してきたメッセージやcreative workです。ブランディングを推進するためには、論理的、戦略的、体系的に考えることはもちろん必須のことです。しかしながら、それだけでは十分ではありません。論理的に考え抜いたその先に、理屈を超えた共感が必要になってきます。

共感を喚起するcreativeの力、そしてそれをアップデートし続けること、これもブランディングには不可欠なものであり、私たちはそこにも大きなエネルギーを注いできました。そうした私たちのブランド再構築の軌跡が、ブランド価値の向上に挑戦されていらっしゃる方々にとってのささやかな示唆となれば、これ以上の喜びはありません。

私たちが紡いできたブランディングという物語に終わりはありません。それはサステナブルでなければなりません。遠く先に視線を送りながら、次代を担う人たちと、これからも物語のつづき、次の夢を描き続けていきます。

When I joined Yokogawa exactly ten years ago, it had customers around the world that were highly influential in the wider society, it had competitive technology, and it had a large number of employees who placed great value on their customers and workplace. From a brand perspective, however, these assets were not consolidated as the company's brand value, and its integrated communications lacked cohesion. And compared to its business scale and the value of the assets I just mentioned, its brand exposure was decisively small.

Increasing the budget for brand exposure is not something that can be done overnight. If that's the case, where then should we start? The first steps should be to consider the brand as a system, consolidate assets that are facing different directions as brand value, and motivate employees to be brand ambassadors.

This is where we started the brand reestablishing project from.

Under the flag of group unity, we engaged our fellow employees, who are the standard-bearers for the brand, and had them consider their social mission and then communicate it. It took ten years to create this framework. This book introduces the steps we took in creating the framework.

This book also features the messages and creative works that we members of the Branding Promotion Team presented both in- and outside Yokogawa while creating the framework. Branding requires theoretical, strategic, and systematic approaches. But that alone is not enough. After you have worked out the theoretical foundation, you must have people identify with your brand in a way that goes beyond logical reasoning.

Branding also needs to have a creative power that can induce identification, which then needs to be continuously updated. We also put a lot of effort into this task. Nothing would please me more than to have the path we took in reestablishing our brand provide a small measure of insight for others who are also engaged in projects to increase brand value.

There is no end to the branding story that we have spun. That is because it must be sustained. While setting our sights on the distant future, we will continue this story together with the next generation and keep sketching our dream for what comes next.

私には夢があります。
宇宙で生み出した技術を
地球の暮らしに応用すること。
植物からプラスチックに代わる素材をつくること。
大切な人のための
オーダーメイドの医療を実現すること——
この星を幸せでいっぱいにすることは、
みんなが夢を語れる世界にすることだと思うから。
私はこれからも夢を見つづけたいと思う。

物語のつづき。夢のはじまり。

この星の物語を紡いでいくために。
この社会をもっと実りあるものにするために。
もっとみなさんと夢の話をしたい。
私たち横河電機は、これからも
美しい夢を見つづける人々を応援していきます。

地球の物語の、つづきを話そう。

YOKOGAWA ◆
Co-innovating tomorrow™

横河電機株式会社

2023年3月に日本経済新聞
に掲載したコーポレートブランド
広告。次の展開を示唆

"The next chapter. It begins
with a dream." ad creative
(newspaper)

未来への共創
Co-innovating tomorrow

横河電機が挑んだ
リブランディングの軌跡

Co-creation for the Future.
The Challenge of Rebranding Yokogawa

2023年10月30日　初版第1刷発行

編著：横河電機ブランドブック制作委員会
編集協力：株式会社日経BPコンサルティング
発行者：林 哲史
発行：株式会社日経BP
発売：株式会社日経BPマーケティング
〒105-8308 東京都港区虎ノ門4-3-12

装丁・本文デザイン：村松 哉 (Dynamite Brothers Syndicate, inc.)
本文DTP：都築 香里（有限会社ビーピーエム）
印刷・製本：図書印刷株式会社

©Nikkei Business Publications, Inc., Nikkei BP Consulting, Inc., 2023
ISBN978-4-296-20306-2 Printed in Japan